SpringerBriefs in Mathematics

SpringerBriefs present concise summaries of cutting-edge research and practical applications across a wide spectrum of fields. Featuring compact volumes of 50 to 125 pages, the series covers a range of content from professional to academic. Briefs are characterized by fast, global electronic dissemination, standard publishing contracts, standardized manuscript preparation and formatting guidelines, and expedited production schedules.

Typical topics might include:

A timely report of state-of-the art techniques A bridge between new research results, as published in journal articles, and a contextual literature review A snapshot of a hot or emerging topic An in-depth case study A presentation of core concepts that students must understand in order to make independent contributions

SpringerBriefs in Mathematics showcases expositions in all areas of mathematics and applied mathematics. Manuscripts presenting new results or a single new result in a classical field, new field, or an emerging topic, applications, or bridges between new results and already published works, are encouraged. The series is intended for mathematicians and applied mathematicians. All works are peer-reviewed to meet the highest standards of scientific literature.

Titles from this series are indexed by Scopus, Web of Science, Mathematical Reviews, and zbMATH.

More information about this series at http://www.springer.com/series/10030

Toshiki Mabuchi

Test Configurations, Stabilities and Canonical Kähler Metrics

Complex Geometry by the Energy Method

 Springer

Toshiki Mabuchi
Department of Mathematics
Osaka University, Graduate School of
Science
Toyonaka, Osaka, Japan

ISSN 2191-8198 ISSN 2191-8201 (electronic)
SpringerBriefs in Mathematics
ISBN 978-981-16-0499-7 ISBN 978-981-16-0500-0 (eBook)
https://doi.org/10.1007/978-981-16-0500-0

This Springer imprint is published by the registered company Springer Nature Singapore Pte Ltd.
The registered company address is: 152 Beach Road, #21-01/04 Gateway East, Singapore 189721,
Singapore

Dedicated to the Memory of
Professor Shoshichi Kobayashi

Preface

This is a development of the lectures on complex geometry delivered at the Yau Mathematical Sciences Center, Tsinghua University during the period 2017–2020. The main purpose of this book is to discuss several topics in complex geometry related to the theory of canonical Kähler metrics for polarized algebraic manifolds.

In 1977, S.-T. Yau solved the Calabi conjecture on the existence of Kähler–Einstein metrics both in the Ricci-flat case and in the Ricci-negative case, where for the Ricci-negative case, independent work was also done by T. Aubin. These allow us to obtain a number of deep consequences in algebraic geometry such as the Miyaoka–Yau inequality and a characterization of the quotients of the 2-dimensional complex unit ball.

For the Ricci-positive case, S.-T. Yau proposed that the existence of Kähler–Einstein metrics on Fano manifolds has some link to the stability in algebraic geometry. It is well-known that G. Tian and S. K. Donaldson introduced the concept of K-stability and K-polystability by slightly different formulations. Now the *Yau–Tian–Donaldson Conjecture* states that:

A Fano manifold X admits a Kähler–Einstein metric

if and only if (X, K_X^{-1}) *is K–polystable.*

Recently, this conjecture was solved affirmatively by X. Chen, S. K. Donaldson, and S. Sun, and independently also by G. Tian. However, in constant scalar curvature Kähler metric cases, this conjecture is still open for general polarizations, or more generally in extremal Kähler cases. In this book, these unsolved cases of this conjecture will also be discussed.

For a Fano manifold, a Kähler–Einstein metric is a typical canonical Kähler metric which plays a very important role in complex geometry. However, such a metric does not necessarily exist on the manifold. In this book, a canonical Kähler metric on a general Fano manifold will also be discussed. Especially, if the Futaki character of a Fano manifold is non-vanishing, the following candidates for canonical Kähler metrics are known: (i) Kähler–Ricci solitons, (ii) extremal

Kähler metrics, (iii) generalized Kähler–Einstein metrics, where the last one is often referred to as M-solitons. In the last chapter, we explain how these three types of metrics differ. Since (i) and (ii) are well-known, we focus on the study of the last one.

Finally, as an appendix, we include a lecture given at Tsinghua University in 2018 in which we discuss how the geometry of L^p-spaces allows us to obtain a natural compactification of the moduli space of graded algebras (such as canonical rings) of a certain type. In contrast to the GIT-limits in algebraic geometry (or to the Gromov–Hausdorff limit in Riemannian geometry), we have some straightforward compactification of the moduli space.

Toyonaka, Japan   Toshiki Mabuchi
January 2020

Contents

Chapter 1
Introduction

Abstract In this chapter, by fixing notation, we introduce various basic concepts to discuss background materials.

- In Sect. 1.1, we introduce several basic concepts such as Kähler classes, polarized algebraic manifolds, CSC Kähler metrics, extremal Kähler metrics and special one-parameter groups.
- In Sect. 1.2, we discuss Deligne pairings with metrics.
- In Sect. 1.3, we give a precise definition of the Chow norm introduced by S. Zhang which plays a very important role in our study of canonical metrics.
- In Sect. 1.4, we give the first and second variation formulas for the Chow norm, and in particular we can show the convexity of the Chow norm along special one-parameter groups.

Keywords Polarized algebraic manifolds · Deligne pairings · The Chow norm

1.1 Preliminaries

For a compact connected complex manifold X, we call $\mathcal{K}$ a *Kähler class* on X if $\mathcal{K}$ is the set of all Kähler forms ω in a fixed Dolbeault cohomology class on X. Here a smooth d-closed $(1, 1)$-form ω on X is called *Kähler* if ω is positive definite everywhere on X. By choosing a system $(z_1, z_2, \ldots, z_n)$ of holomorphic local coordinates on X, we can write a Kähler form ω on X as

$$\omega = \frac{\sqrt{-1}}{2\pi} \sum_{\alpha, \beta} g_{\alpha\bar{\beta}} \, dz_\alpha \wedge dz_{\bar{\beta}}, \tag{1.1}$$

where $d\omega = 0$. The Ricci form $\mathrm{Ric}(\omega)$ for ω is

$$\mathrm{Ric}(\omega) := \frac{\sqrt{-1}}{2\pi} \sum_{\alpha, \beta=1}^{n} R_{\alpha\bar{\beta}} \, dz_\alpha \wedge dz_{\bar{\beta}} = -dd^c \log \omega^n, \tag{1.2}$$

T. Mabuchi, *Test Configurations, Stabilities and Canonical Kähler Metrics*,
SpringerBriefs in Mathematics, https://doi.org/10.1007/978-981-16-0500-0_1

where $2\pi dd^c := \sqrt{-1}\partial\bar{\partial}$. The scalar curvature S_ω for ω is

$$S_\omega := \mathrm{Tr}_\omega \mathrm{Ric}(\omega) = \sum_{\alpha,\beta=1}^{n} g^{\bar{\beta}\alpha} R_{\alpha\bar{\beta}}.$$

Here $(g^{\bar{\beta}\alpha})$ is the inverse matrix of $(g_{\alpha\bar{\beta}})$. Put $Z_\omega := \mathrm{grad}_\omega^{\mathbb{C}} S_\omega$, where for each complex-valued smooth function ψ on X, we define a complex vector field

$$\mathrm{grad}_\omega^{\mathbb{C}} \psi = \frac{1}{\sqrt{-1}} \sum_{\alpha,\beta} g^{\bar{\beta}\alpha} \frac{\partial \psi}{\partial z_{\bar{\beta}}} \frac{\partial}{\partial z_\alpha} \tag{1.3}$$

of type $(1,0)$ on X. Then ω is called *CSC Kähler* if S_ω is a constant function. More generally, if the vector field Z_ω is holomorphic, then ω is called *extremal Kähler* (cf. [8, 9]) and in such a case, Z_ω is called the *extremal vector field* for (X, ω).

Definition 1.1 (X, L) is called a *polarized algebraic manifold* if L is an ample line bundle on a compact complex connected manifold X. However, unless otherwise stated, L is always assumed to be very ample (with the only exception for $L = K_X^{-1}$).

For a Hermitian vector space V of complex dimension N, let $\sigma : \mathbb{C}^* \to \mathrm{GL}(V)$ be an algebraic group homomorphism. Then for a suitable choice of an orthonormal basis for V, we can diagonalize σ in the form

$$\sigma(t) = \begin{pmatrix} t^{\alpha_1} & & \text{\Large 0} \\ & t^{\alpha_2} & \\ & & \ddots \\ \text{\Large 0} & & t^{\alpha_N} \end{pmatrix}, \qquad t \in \mathbb{C}^*,$$

where the integers $\alpha_1, \alpha_2, \ldots, \alpha_N$ independent of t are called the *weights* of the $\mathbb{C}^*$-action on V via σ. Let F be the finite cyclic subgroup of $\mathrm{SL}(V)$ defined by

$$F := \{ \zeta \, \mathrm{id}_V \, ; \, \zeta \in \mathbb{C} \text{ is an } N\text{-th root of unity} \}.$$

By setting $\alpha_0 := (\alpha_1 + \alpha_2 + \cdots + \alpha_N)/N$, we have rational numbers $\bar{\alpha}_i := \alpha_i - \alpha_0$. Then by setting $\sigma^{\mathrm{SL}}(t) := \{\det \sigma(t)\}^{-1/N} \sigma(t)$, we have an algebraic group homomorphism $\sigma^{\mathrm{SL}} : \mathbb{C}^* \to \mathrm{SL}(V)/F$, called the *special linearization of* σ, written as

$$\sigma^{\mathrm{SL}}(t) = \begin{pmatrix} t^{\bar{\alpha}_1} & & \text{\Large 0} \\ & t^{\bar{\alpha}_2} & \\ & & \ddots \\ \text{\Large 0} & & t^{\bar{\alpha}_N} \end{pmatrix}, \qquad t \in \mathbb{C}^*,$$

modulo F. Let $\mathbb{R}_+$ be the multiplicative group of positive real numbers. Then σ^{SL}, when restricted to $\mathbb{R}_+$, defines a homomorphism: $\mathbb{R}_+ \to SL(V)$ such that the eigenvalues of $\sigma^{SL}(t)$, $t \in \mathbb{R}_+$, are all positive real numbers. Then $\bar{\alpha}_1, \bar{\alpha}_2, \cdots,$ $\bar{\alpha}_N$ are called the *weights of the $\mathbb{R}_+$-action on V via σ^{SL}*. Moreover for σ and σ^{SL},

$$\begin{pmatrix} \alpha_1 & & 0 \\ & \alpha_2 & \\ & & \cdots \\ 0 & & \alpha_N \end{pmatrix}, \quad \begin{pmatrix} \bar{\alpha}_1 & & 0 \\ & \bar{\alpha}_2 & \\ & & \cdots \\ 0 & & \bar{\alpha}_N \end{pmatrix}$$

in $\mathfrak{sl}(V)$ are called the *fundamental generators of σ, σ^{SL}*, respectively.

Let $\mathbb{A}^1 := \{ z \in \mathbb{C} \}$ be a complex affine line on which $\mathbb{C}^*$ acts by multiplication of complex numbers. Let $\pi : E \to \mathbb{A}^1$ be an algebraic vector bundle over $\mathbb{A}^1$ such that $\mathbb{C}^*$ acts on E covering the $\mathbb{C}^*$-action on $\mathbb{A}^1$ and inducing linear maps between fibers of E. Let ρ be a Hermitian metric for the fiber E_1 over $1 \in \mathbb{A}^1$. Since the origin in $\mathbb{A}^1$ is fixed by the $\mathbb{C}^*$-action, the fiber E_0 over the origin is preserved by the $\mathbb{C}^*$-action, so that the $\mathbb{C}^*$-action on E_0 is given by a representation

$$\sigma : \ \mathbb{C}^* \to GL(E_0).$$

Then the affirmative solution [56] of the *equivariant Serre conjecture for abelian groups* allows us to trivialize E equivariantly in the form

$$E \cong E_0 \times \mathbb{A}^1,$$

where $\mathbb{C}^*$ acts on $E_0 \times \mathbb{A}^1$ by

$$\mathbb{C}^* \times (E_0 \times \mathbb{A}^1) \to E_0 \times \mathbb{A}^1, \quad (t, (e, z)) \mapsto (\sigma(t)e, tz).$$

Then by Lemma 2 in Donaldson [21], a trivialization as above can be chosen isometrically in such a way that the Hermitian metric ρ on E_1 induces a Hermitian metric for the fiber E_0 on which S^1 in $\mathbb{C}^*$ acts isometrically.

1.2 The Deligne Pairings with Metrics

Let $\pi : Y \to T$ be a flat projective morphism of irreducible complex varieties of relative dimension $n := \dim Y - \dim T \geq 0$ such that T is smooth. Here a *complex variety* means a reduced (possibly reducible) algebraic variety defined over $\mathbb{C}$. Let

$$\mathscr{L}_i = \mathcal{O}_Y(D_i), \quad i = 0, 1, \ldots, n,$$

be line bundles over Y, where D_i is a Cartier divisor on Y. Then the intersection

$$\delta := D_0 \cdot D_1 \cdots D_n$$

is a q-dimensional algebraic cycle on Y with $q = \dim Y - (n+1) = \dim T - 1$. Since T is smooth, the pushforward $\pi_* \delta$ defines a line bundle on T,

$$\langle \mathscr{L}_0, \mathscr{L}_1, \dots, \mathscr{L}_n \rangle_{Y/T},$$

called the *Deligne pairing* of $\mathscr{L}_0, \mathscr{L}_1, \dots, \mathscr{L}_n$. Let ℓ_i, $i = 0, 1, \dots, n$, be local (with respect to T) sections for $\mathscr{L}_i$ such that the corresponding divisors have an empty intersection. Then the symbol

$$\langle \ell_0, \ell_1, \dots, \ell_n \rangle$$

represents a local basis for $\mathscr{L} := \langle \mathscr{L}_0, \mathscr{L}_1, \dots, \mathscr{L}_n \rangle_{Y/T}$. For a Hermitian metric h_i for the line bundle $\mathscr{L}_i$, we can write

$$h_i = e^{-\phi_i}, \qquad i = 0, 1, \dots, n,$$

by using the associated Kähler potential $\phi_i := -\log h_i$. Then the Deligne pairing $\mathscr{L}$ has a Hermitian metric $e^{-\phi}$ with a continuous Kähler potential

$$\phi = \langle \phi_0, \phi_1, \dots, \phi_n \rangle_{Y/T},$$

smooth over the smooth locus of π, which is multilinear and characterized by the following properties:

$$dd^c \langle \phi_0, \phi_1, \dots, \phi_n \rangle_{Y/T} = \int_{Y/T} \prod_{i=0}^{n} dd^c \phi_i,$$

$$\tag{1.4}$$

$$\langle \phi_0', \phi_1, \dots, \phi_n \rangle_{Y/T} - \langle \phi_0'', \phi_1, \dots, \phi_n \rangle_{Y/T} = \int_{Y/T} (\phi_0' - \phi_0'') \prod_{i=1}^{n} dd^c \phi_i,$$

$$\tag{1.5}$$

where $\int_{Y/T}$ denotes the fiber integration for the fiber space Y sitting over T. For more details, see [5, 60, 68, 93].

Remark 1.1 (cf. [68, 93]) For $D_0 := \mathrm{zero}(\ell_0)$, by choosing D_0 as a difference of irreducible reduced divisors, it is possible to define $\phi = \langle \phi_0, \phi_1, \dots, \phi_n \rangle_{Y/T}$ by induction on n by the formula

$$\phi(\ell) = \phi'(\ell') + \int_{Y/T} \phi_0(\ell_0) \prod_{i=1}^{n} dd^c \phi_i,$$

where for $\phi = \langle \phi_0, \phi_1, \ldots, \phi_n \rangle_{Y/T}$, $\phi' = \langle \phi_1, \ldots, \phi_n \rangle_{D_0/T}$, $\ell = \langle \ell_0, \ell_1, \ldots, \ell_n \rangle$, $\ell' = \langle \ell_1, \ldots, \ell_n \rangle$, $h_\phi = e^{-\phi}$, $h_{\phi'} = e^{-\phi'}$, $h_0 = e^{-\phi_0}$, we put

$$\phi(\ell) = -\log h_\phi(\ell, \ell), \quad \phi'(\ell') = -\log h_{\phi'}(\ell', \ell'), \quad \phi_0(\ell_0) = -\log h_0(\ell_0, \ell_0).$$

Note also that, for $n = 0$, $\langle \mathscr{L}_0 \rangle_{Y/T}$ is nothing but the norm of $\mathscr{L}_0$ with respect to the finite flat morphism $\pi : Y \to T$, and in particular for $\ell = \langle \ell_0 \rangle$,

$$\{\phi(\ell)\}(t) = \sum_{P \in \pi^{-1}(t)} \{\phi_0(\ell_0)\}(P), \quad t \in T.$$

Remark 1.2 (cf. [68, 93]) For bases $\langle \ell_0, \ell_1, \cdots, \ell_n \rangle$, $\langle \ell'_0, \ell'_1, \cdots, \ell'_n \rangle$ over an open subset of T, the transition function is defined as follows: It suffices to consider the case where, for some k, we have $\ell_i = \ell'_i$ for all $i \neq k$. For a point t in the open set, let

$$Y_t \cap \left\{ \bigcap_{i \neq k} \text{zero}(\ell_i) \right\} = \sum_\alpha n_\alpha(t) P_\alpha(t)$$

be a zero cycle on $Y_t := \pi^{-1}(t)$, where for each t, $n_\alpha(t)$ is an integer, and $\{P_\alpha(t)\}$ is a finite set of points in Y_t. Then by setting $f := \ell'_k / \ell_k$, we have

$$\langle \ell'_0, \ell'_1, \cdots, \ell'_n \rangle = \left\{ \prod_\alpha f(P_\alpha(t))^{n_\alpha(t)} \right\} \langle \ell_0, \ell_1, \cdots, \ell_n \rangle.$$

1.3 Definition of the Chow Norm

For $V = \mathbb{C}^N$ with the standard Euclidean norm ρ, we consider its dual $V^* = \mathbb{C}^N = \{z = (z_1, \ldots, z_N)\}$. Then the group $SL(V) = SL(N, \mathbb{C})$ acts on V^* by the contragradient representation. For $y = (y_1, \ldots, y_N) \in V$ and $z = (z_1, \ldots, z_N) \in V^*$, we put

$$|y|^2 := \sum_{i=1}^N |y_i|^2 \quad \text{and} \quad |z|^2 = \sum_{i=1}^N |z_i|^2.$$

Furthermore, for $y = (y^{(0)}, y^{(1)}, \cdots, y^{(n)}) \in V^{n+1}$, we put

$$\|y\|^2 := \prod_{j=0}^n |y^{(j)}|^2,$$

and let $\omega_{\text{FS}}([y^{(j)}])$ be the Fubini–Study form $dd^c \log |y^{(j)}|^2$ for the projective space $\mathbb{P}(V) = \{[y^{(j)}]; 0 \neq y^{(j)} \in \mathbb{C}^N\}$, $j = 0, 1, \ldots, n$. Put $W := (\text{Sym}^d V^*)^{\otimes n+1}$ for a positive integer d. Then for every $w \in W$, we define its *Chow norm* $\|w\|_{\text{CH}(\rho)} \geq 0$ by

$$\|w\|_{\text{CH}(\rho)} := \exp\left\{ \int_{\mathbb{P}(V)^{n+1}} \left(\log \frac{|w(y)|}{\|y\|^d} \right) \prod_{j=0}^{n} \omega_{\text{FS}}^{N-1}([y^{(j)}]) \right\},$$

where $w = w(y)$ is viewed as a homogeneous polynomial in $(y^{(0)}, y^{(1)}, \cdots, y^{(n)})$ of multi-degree $(d, d, \cdots, d)$. Obviously, $\| \ \|_{\text{CH}(\rho)}$ defines a Finsler norm on W. For more details, see [70, 93].

1.4 The First and Second Variation Formulas for the Chow Norm

Let $X \subset \mathbb{P}(V^*)$ be an n-dimensional projective variety for $V^* = \{z = (z_0, \ldots, z_N)\}$ as in the preceding section, and we consider the Fubini–Study form

$$\omega_{\text{FS}} = dd^c \log |z|^2 = dd^c \log \sum_{i=0}^{N} |z_i|^2$$

on $\mathbb{P}(V^*)$. Put $d := \deg_{\mathbb{P}(V^*)} X$ and $W := (\text{Sym}^d V^*)^{\otimes n+1}$. Let $0 \neq \hat{X} \in W$ be the Chow form for X, so that the corresponding point $[\hat{X}]$ in $\mathbb{P}(W)$ is the Chow point for X viewed as an algebraic cycle on $\mathbb{P}(V^*)$. Let

$$\sigma : \mathbb{C}^* \to \text{SL}(V)$$

be an algebraic group homomorphism, where $\text{SL}(V)$ acts on V^* by contragradient representation. (More generally, the arguments down below are valid also for a Lie group homomorphism σ from $\mathbb{R}_+$ to $\text{SL}(V)$ with rational weights $-b_1, \cdots, -b_N$.) Since $\mathscr{O}_{\mathbb{P}(V^*)}(-1) \setminus \{\text{zero section}\} = V^* \setminus \{0\}$, the mapping

$$V^* \setminus \{0\} \ni z \mapsto |\sigma(t)z|^2 \in \mathbb{R}$$

defines a Hermitian metric on the line bundle $\mathscr{O}_{\mathbb{P}(V^*)}(-1)$ for each fixed $t \in \mathbb{C}^*$. Then by taking its dual, we can view

$$\phi_t = \log |\sigma(t)z|^2 \tag{1.6}$$

as a Kähler potential for the polarization class $c_1(\mathcal{O}_{\mathbb{P}(V^*)}(1))$ on $\mathbb{P}(V^*)$. Note that $\sigma(t)^*\omega_{FS} = dd^c\phi_t$. Put $\dot{\phi}_t := \partial\phi_t/\partial t$. Then (see for instance [70]):

$$\frac{d}{dt}\log\|\sigma(t)\cdot\hat{X}\|_{CH(\rho)} = \frac{n+1}{2}\int_X \dot{\phi}_t(dd^c\phi_t)^n, \quad (1.7)$$

where $SL(V)$ acts on $W = (\text{Sym}^d V^*)^{\otimes n+1}$ induced by the contragradient representation of $SL(V)$ on V^*. For a suitable basis for V, we can diagonalize $\sigma(t)$ as

$$\sigma(t) = \begin{pmatrix} t^{-b_1} & & & 0 \\ & t^{-b_2} & & \\ & & \cdots & \\ 0 & & & t^{-b_N} \end{pmatrix}, \quad t \in \mathbb{C}^*,$$

where $-b_\alpha \in \mathbb{Z}$, $\alpha = 1, 2, \cdots, N$, are the weights of the $\mathbb{C}^*$-action on V via σ. Then by the contragradient representation, we can write

$$\sigma(t)z = (t^{b_1}z_1, t^{b_2}z_2, \ldots, t^{b_N}z_N), \quad z \in V^*.$$

Put $f(s) := \log\|\sigma(t)\cdot\hat{X}\|_{CH(\rho)}$, where $t = \exp s$ for $s \in \mathbb{R}$. Then by (1.6),

$$\phi_t = \log(e^{2sb_1}|z_1|^2 + \cdots + e^{2sb_N}|z_N|^2).$$

Note that $\omega_{FS} = dd^c\log(|z_1|^2 + \cdots + |z_N|^2)$. Put $\dot{f}(s) = \partial f/\partial s$. Then by (1.7), we have the *first variation formula for the Chow norm*:

$$\dot{f}(0) = (n+1)\int_X \frac{b_1|z_1|^2 + \cdots + b_N|z_N|^2}{|z_1|^2 + \cdots + |z_N|^2} \omega_{FS}^n. \quad (1.8)$$

Consider the rectangle $R_{a,b} = \{\zeta \in \mathbb{C}; a \leq \text{Re }\zeta \leq b, 0 \leq \text{Im }\zeta \leq 2\pi\}$, where real numbers a, b are such that $a < b$. Let $\eta : X \times R_{a,b} \to \mathbb{P}(V^*)$ be the map defined by

$$\eta(x, \zeta) := \sigma(\exp\zeta)\cdot x.$$

Then we have the following *second variation formula for the Chow norm*:

$$\dot{f}(b) - \dot{f}(a) = \int_{X\times R_{a,b}} \eta^*\omega_{FS}^{n+1} \geq 0. \quad (1.9)$$

In particular, $f(s)$ is a convex function of s, i.e., $\ddot{f}(s) \geq 0$ for all s, where $\ddot{f}(s) := \partial^2 f/\partial s^2$. For more details, see for instance [48, 70, 93].

Problems

1.1 Show that, if $\ddot{f}(0) = 0$ for $f(s)$ above, then the one-parameter group $\sigma : \mathbb{C}^* \to$ SL(V) preserves the subvariety X in $\mathbb{P}(V^*)$.

1.2 For a smooth irreducible projective variety X with positive first Chern class, let ω be a CSC Kähler metric in the polarization class $c_1(X)$. Show that Ric$(\omega) = \omega$.

Chapter 2
The Donaldson–Futaki Invariant

Abstract In the study of the Calabi conjecture for Fano manifolds, G. Tian introduced the concept of special degenerations to characterize the stability related to the existence of Kähler–Einstein metrics. Later, S.K. Donaldson reformulated this concept to obtain a purely algebraic concept of test configurations.

- In Sects. 2.1 and 2.2, we define the concept of test configurations and study its elementary properties.
- In Sect. 2.3, we define Donaldson–Futaki invariants DF_i for test configurations.
- In Sect. 2.4, the Donaldson–Futaki invariant DF_1 for a test configuration is characterized as an intersection number.
- In Sect. 2.5, the Chow weight (the derivative of the Chow norm at infinity) along a special one-parameter group for a test configuration is expressed in terms of the Donaldson–Futaki invariants.

Keywords Test configurations · The Donaldson–Futaki invariant

2.1 Test Configurations

Let (X, L) be a polarized algebraic manifold. Let $\mathbb{C}^*$ act on the complex affine line $\mathbb{A}^1 := \{z \in \mathbb{C}\}$ by multiplication of complex numbers:

$$\mathbb{C}^* \times \mathbb{A}^1 \ni (t, x) \mapsto tx \in \mathbb{A}^1.$$

A pair $(\mathscr{X}, \mathscr{L})$ is called a *test configuration* ([20]; see also [79]) for (X, L) if $\mathbb{C}^*$ acts on $\mathscr{X}$ such that there exists a $\mathbb{C}^*$-equivariant projective morphism

$$\pi : \mathscr{X} \to \mathbb{A}^1$$

of complex varieties satisfying the following conditions:

- $\mathscr{L}$ is a relatively very ample line bundle on the fiber space $\mathscr{X}$ over $\mathbb{A}^1$ such that the $\mathbb{C}^*$-action on $\mathscr{X}$ lifts to an action on $\mathscr{L}$ inducing linear maps between fibers.

T. Mabuchi, *Test Configurations, Stabilities and Canonical Kähler Metrics*,
SpringerBriefs in Mathematics, https://doi.org/10.1007/978-981-16-0500-0_2

- For some positive integer γ, there exists an isomorphism $(\mathscr{X}_1, \mathscr{L}_1) \cong (X, L^{\otimes \gamma})$, where $(\mathscr{X}_z, \mathscr{L}_z)$ denotes the fiber of $(\mathscr{X}, \mathscr{L})$ over each $z \in \mathbb{A}^1$.

Here γ is called the *exponent* of the test configuration $(\mathscr{X}, \mathscr{L})$ for (X, L). The action of each $t \in \mathbb{C}^*$ on $\mathscr{X}$ (resp. $\mathscr{L}$) is written as

$$g(t) \in \operatorname{Aut}(\mathscr{X}) \quad (\text{resp. } \tilde{g}(t) \in \operatorname{Aut}(\mathscr{L})),$$

where $\operatorname{Aut}(\mathscr{X})$ (resp. $\operatorname{Aut}(\mathscr{L})$) is the set of all holomorphic automorphisms of $\mathscr{X}$ (resp. $\mathscr{L}$) viewed just as a complex variety. We now put $X_{\mathbb{A}^1 \setminus \{0\}} := X \times (\mathbb{A}^1 \setminus \{0\})$. Then by identifying $\mathscr{X}_1$ with X, we have the $\mathbb{C}^*$-equivariant isomorphism

$$X_{\mathbb{A}^1 \setminus \{0\}} = \mathscr{X}_1 \times (\mathbb{A}^1 \setminus \{0\}) \cong \mathscr{X} \setminus \mathscr{X}_0 \tag{2.1}$$

$$(x, t) \quad \longleftrightarrow \quad g(t) \cdot x,$$

where $\mathbb{C}^*$ acts on $X_{\mathbb{A}^1 \setminus \{0\}} = X \times (\mathbb{A}^1 \setminus \{0\})$ by multiplication of complex numbers just on the second factor $(\mathbb{A}^1 \setminus \{0\})$. Now by attaching an end $X \times \{\infty\}$ to $X_{\mathbb{A}^1 \setminus \{0\}}$, we have a $\mathbb{C}^*$-equivariant closure $X_{\mathbb{P}^1 \setminus \{0\}} := X \times (\mathbb{P}^1 \setminus \{0\})$ of $X_{\mathbb{A}^1 \setminus \{0\}}$. Then by gluing $\mathscr{X}$ and $X_{\mathbb{P}^1 \setminus \{0\}}$ together, we have the $\mathbb{C}^*$-equivariant compactification

$$\bar{\mathscr{X}} := X_{\mathbb{P}^1 \setminus \{0\}} \cup \mathscr{X}$$

of $\mathscr{X}$ via the identification (2.1) such that the $\mathbb{C}^*$-action fixes the infinity fiber $\bar{\mathscr{X}}_\infty$. Similarly, via the $\mathbb{C}^*$-equivariant identification

$$\mathscr{L} \setminus \mathscr{L}_0 \cong L^{\otimes \gamma}_{\mathbb{A}^1 \setminus \{0\}} := \mathscr{L}_1 \times (\mathbb{A}^1 \setminus \{0\})$$

$$\tilde{g}(t) \cdot \ell \quad \longleftrightarrow \quad (\ell, t),$$

we can glue $\mathscr{L}$ and $L^{\otimes \gamma}_{\mathbb{P}^1 \setminus \{0\}} := \mathscr{L}_1 \times (\mathbb{P}^1 \setminus \{0\})$ together to obtain

$$\bar{\mathscr{L}} := L^{\otimes \gamma}_{\mathbb{P}^1 \setminus \{0\}} \cup \mathscr{L}.$$

on which the group $\mathbb{C}^*$ acts fixing the infinity fiber $\bar{\mathscr{L}}_\infty$. Obviously, $\bar{\mathscr{L}}$ is relatively very ample on the fiber space $\bar{\mathscr{X}}$ over $\mathbb{P}^1$. Thus we obtain the compactified family

$$(\bar{\mathscr{X}}, \bar{\mathscr{L}}) = \bigcup_{z \in \mathbb{P}^1} (\bar{\mathscr{X}}_z, \bar{\mathscr{L}}_z),$$

where the restriction of $\bar{\mathscr{L}}$ over $\mathbb{P}^1 \setminus \{0\}$ is trivialized as the bundle $L^{\otimes \gamma}_{\mathbb{P}^1 \setminus \{0\}}$.

- A test configuration $(\mathscr{X}, \mathscr{L})$ for (X, L) is called a *product configuration* if $\mathscr{X}$ over $\mathbb{A}^1$ is isomorphic to the product $X \times \mathbb{A}^1$ just as a complex variety.

- A test configuration $(\mathscr{X}, \mathscr{L})$ for (X, L) is called *trivial* if $(\mathscr{X}, \mathscr{L})$ is a product configuration and furthermore $\mathbb{C}^*$ acts trivially on the first factor of $\mathscr{X} \cong X \times \mathbb{A}^1$.

A test configuration

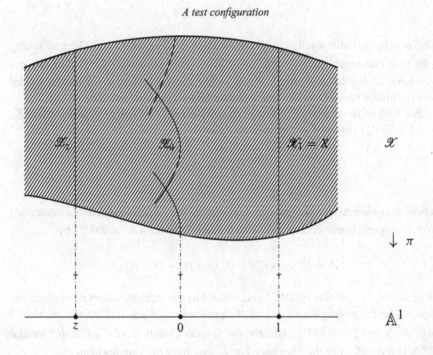

2.2 Test Configurations Associated to One-Parameter Groups

For a polarized algebraic manifold (X, L), we put $V_\gamma := H^0(X, L^{\otimes\gamma})$ for each positive integer γ. Fix a Hermitian metric h for L such that

$$\omega := \text{Ric}(h) = -dd^c \log h$$

is Kähler. Let ρ_γ be the Hermitian structure for V_γ induced by h and ω such that

$$\langle v_1, v_2 \rangle_{\rho_\gamma} := \int_X (v_1, v_2)_h \, \omega^n, \qquad v_1, v_2 \in V_\gamma, \tag{2.2}$$

where $(\ ,\ )_h$ is the pointwise Hermitian pairing for L by the Hermitian metric h. Let S^1 be the maximal compact subgroup of $\mathbb{C}^*$. An algebraic group homomorphism

$$\sigma : \mathbb{C}^* \rightarrow \text{GL}(V_\gamma)$$

is called a *special one-parameter group* if $\sigma(S^1)$ acts isometrically on (V_γ, ρ_γ). In view of Sect. 1.1, we may choose another approach by assuming that σ is an algebraic group homomorphism:

$$\sigma : \mathbb{C}^* \to \mathrm{SL}(V_\gamma)/F,$$

where F is the finite cyclic subgroup of $\mathrm{SL}(V_\gamma)$ as in Sect. 1.1. Then by taking a finite cyclic unramified cover $\tilde{\mathbb{C}}^*$ of $\mathbb{C}^*$, we have a lift $\tilde\sigma : \tilde{\mathbb{C}}^* \to \mathrm{SL}(V_\gamma)$ of σ, where $\tilde{\mathbb{C}}^*$ is also an algebraic torus. In this approach also, σ is called a special one-parameter group if $\tilde\sigma(S^1)$ acts isometrically on (V_γ, ρ_γ).

For both of these approaches, we can consider a $\mathbb{C}^*$-invariant closed subset $\mathscr{X}_\sigma$ of $\mathbb{A}^1 \times \mathbb{P}(V_\gamma^*)$ obtained as the closure of the set

$$\bigcup_{t \in \mathbb{C}^*} \{t\} \times \sigma(t)X,$$

where X is viewed as a subvariety of $\mathbb{P}(V_\gamma^*)$ by the Kodaira embedding associated to the complete linear system $|L^{\otimes\gamma}|$ on X, and $\mathbb{C}^*$ acts on $\mathbb{A}^1 \times \mathbb{P}(V_\gamma^*)$ by

$$\mathbb{C}^* \times \mathbb{A}^1 \times \mathbb{P}(V_\gamma^*) \ni (t, (z, x)) \mapsto (tz, \sigma(t)x).$$

Here the $\mathrm{GL}(V_\gamma)$-action on $\mathbb{P}(V_\gamma^*)$ is induced by the contragradient representation. Let $\mathscr{L}_\sigma$ be the restriction to $\mathscr{X}_\sigma$ of the pullback $\mathrm{pr}_2^* \, \mathcal{O}_{\mathbb{P}(V_\gamma^*)}(1)$ by the projection $\mathrm{pr}_2 : \mathbb{A}^1 \times \mathbb{P}(V_\gamma^*) \to \mathbb{P}(V_\gamma^*)$. Clearly, the natural $\mathbb{C}^*$-action on $\mathscr{X}_\sigma$ lifts to a natural $\mathbb{C}^*$-action on $\mathscr{L}_\sigma$. For the fiber over $1 \in \mathbb{A}^1$, we have the isomorphism

$$((\mathscr{X}_\sigma)_1, (\mathscr{L}_\sigma)_1) \cong (X, L^{\otimes\gamma}).$$

- Hence $(\mathscr{X}_\sigma, \mathscr{L}_\sigma)$ is a test configuration of exponent γ for (X, L), and is called the *test configuration associated to the special one-parameter group* σ.
- $(\mathscr{X}_\sigma, \mathscr{L}_\sigma)$ is called *trivial* if σ^{SL} is a trivial homomorphism.

Let $(\mathscr{X}, \mathscr{L})$ be a test configuration, of exponent γ, for (X, L) as in Sect. 2.1. Then the direct image sheaf

$$E := \pi_* \mathscr{L}^{\otimes m}$$

is viewed as a vector bundle over $\mathbb{A}^1$ on which $\mathbb{C}^*$ acts. In view of Sect. 1.1, the vector bundle E can be trivialized $\mathbb{C}^*$-equivariantly as

$$E \cong E_0 \times \mathbb{A}^1,$$

where for each positive integer m, the algebraic torus $\mathbb{C}^*$ acts on the fiber E_0 of E over the origin via a representation

$$\sigma_m : \mathbb{C}^* \to \mathrm{GL}(E_0).$$

Put $\gamma_m := m\gamma$. By the identification of E_0 with $E_1 = H^0(X, L^{\otimes \gamma_m}) = V_{\gamma_m}$, the compact subgroup S^1 in $\mathbb{C}^*$ acts isometrically on V_{γ_m} with the Hermitian structure ρ_{γ_m}. In particular, σ_m above is a special one-parameter group, and we can write

$$(\mathscr{X}, \mathscr{L}^{\otimes m}) = (\mathscr{X}_{\sigma_m}, \mathscr{L}_{\sigma_m}), \qquad m = 1, 2, \cdots.$$

2.3   Definition of the Donaldson–Futaki Invariant

Let (X, L) be a polarized algebraic manifold. We then consider a test configuration $(\mathscr{X}, \mathscr{L})$ for (X, L) of exponent γ. For a sufficiently large integer $m \gg 1$, we put

$$N_m := \dim H^0(X, L^{\otimes m\gamma}) = \dim H^0(\mathscr{X}_0, \mathscr{L}_0^{\otimes m}),$$

$$w_m := \text{weight of the } \mathbb{C}^*\text{-action on } \det H^0(\mathscr{X}_0, \mathscr{L}_0^{\otimes m}).$$

Then by the Riemann–Roch theorem and also by the equivariant Riemann–Roch theorem, we can write N_m and w_m as polynomials with rational coefficients,

$$N_m = a_0 m^n + a_1 m^{n-1} + \cdots + a_{n-1} m + a_n,$$

$$w_m = b_0 m^{n+1} + b_1 m^n + \cdots + b_n m + b_{n+1},$$

where $a_0 = \gamma^n c_1(L)^n[X]/n!$ is positive. Hence we have the asymptotic expansion

$$\frac{w_m}{mN_m} = \frac{b_0 m^{n+1} + b_1 m^n + \cdots}{a_0 m^{n+1} + a_1 m^n + \cdots} = \frac{b_0}{a_0} + \frac{a_0 b_1 - a_1 b_0}{a_0^2} m^{-1} + \cdots$$

$$= \sum_{i=0}^{\infty} \mathrm{DF}_i(\mathscr{X}, \mathscr{L}) m^{-i}.$$

From this expansion, we immediately obtain

$$\mathrm{DF}_0(\mathscr{X}, \mathscr{L}) = \frac{b_0}{a_0}, \qquad \mathrm{DF}_1(\mathscr{X}, \mathscr{L}) = \frac{a_0 b_1 - a_1 b_0}{a_0^2}. \tag{2.3}$$

Definition 2.1 $\mathrm{DF}_1(\mathscr{X}, \mathscr{L})$ is called the *Donaldson–Futaki invariant* (see [20]) of the test configuration $(\mathscr{X}, \mathscr{L})$.

Remark 2.1 If we replace $\mathscr{L}$ by $\mathscr{L}^{\otimes k}$ for some positive integer k, then the terms a_0, a_1, b_0, b_1 are replaced by $a_0 k^n, a_1 k^{n-1}, b_0 k^{n+1}, b_1 k^n$, respectively. Hence by (2.3), we see that $\mathrm{DF}_1(\mathscr{X}, \mathscr{L}^{\otimes k}) = \mathrm{DF}_1(\mathscr{X}, \mathscr{L})$.

2.4 Expression of DF$_1$ as an Intersection Number

For a test configuration $(\mathscr{X}, \mathscr{L})$ of exponent γ for (X, L), we assume that $\mathscr{X}$ is a normal complex variety. Let us consider the relative canonical divisor

$$K_{\bar{\mathscr{X}}/\mathbb{P}^1} := K_{\bar{\mathscr{X}}} - \bar{\pi}^* K_{\mathbb{P}^1},$$

where $K_{\bar{\mathscr{X}}/\mathbb{P}^1}$ is a Weil divisor on $\bar{\mathscr{X}}$, and $\bar{\pi} : \bar{\mathscr{X}} \to \mathbb{P}^1$ is the natural projection. The purpose of this section is to prove the following formula ([65, 82]; see also [4]):

Theorem 2.1

$$\mathrm{DF}_1(\mathscr{X}, \mathscr{L}) = -\frac{1}{2\mathrm{vol}} \left\{ (K_{\bar{\mathscr{X}}/\mathbb{P}^1} \cdot \bar{\mathscr{L}}^n) + \frac{\bar{S}}{n+1}(\bar{\mathscr{L}}^{n+1}) \right\},$$

where $\mathrm{vol} = \gamma^n c_1(L)^n[X]$ and $\bar{S} := n \, \mathrm{vol}^{-1} c_1(X) \gamma^{n-1} c_1(L)^{n-1}[X]$, and the intersection numbers $(K_{\bar{\mathscr{X}}/\mathbb{P}^1} \cdot \bar{\mathscr{L}}^n)$, $(\bar{\mathscr{L}}^{n+1})$ are taken on $\bar{\mathscr{X}}$.

Proof Let $m \gg 1$. Note that the direct image sheaf

$$\mathscr{E}^{(m)} := \bar{\pi}_* \bar{\mathscr{L}}^{\otimes m}$$

over $\mathbb{P}^1$ is locally free of rank N_m. We then consider the line bundle

$$\mathscr{N} := \det \mathscr{E}^{(m)} \qquad (2.4)$$

on $\mathbb{P}^1$. By viewing $\mathscr{N}$ just as a complex variety, we consider the group $Aut(\mathscr{N})$ of all holomorphic automorphisms of $\mathscr{N}$. Then the $\mathbb{C}^*$-action on $\mathscr{N}$ induced by the $\mathbb{C}^*$-action on $\mathscr{L}$ is expressible as

$$\mathbb{C}^* \times \mathscr{N} \ni (t, v) \mapsto g(t)v \in \mathscr{N},$$

for some group homomorphism $g : \mathbb{C}^* \to Aut(\mathscr{N})$. Actually, $g(t), t \in \mathbb{C}^*$, define bundle automorphisms of $\mathscr{N}$ covering the $\mathbb{C}^*$-action on $\mathbb{P}^1$. We now choose an element $0 \neq v_1 \in \mathscr{N}_1$ sitting over $1 \in \mathbb{A}^1$. Then

$$\sigma(t) := g(t)v_1, \qquad t \in \mathbb{C}^* (= \mathbb{A}^1 \setminus \{0\}),$$

extends to a rational section of $\mathcal{N}$ over $\mathbb{P}^1$, written also as

$$\sigma = \sigma(z), \qquad z \in \mathbb{P}^1,$$

which is holomorphic and nowhere vanishing when restricted to $\mathbb{P}^1 \setminus \{0\}$. Then the order $\mu := \mathrm{ord}_{z=0}\,\sigma(z)$ of σ at the origin is nothing but $\deg \mathcal{N}$. Hence

$$\nu_0 := \lim_{z \to 0} z^{-\mu}\sigma(z)$$

is a nonzero element of $\mathcal{N}_0$. Then for every $t \in \mathbb{C}^*$,

$$g(t)\nu_0 = g(t)\left\{\lim_{z \to 0} z^{-\mu}\sigma(z)\right\} = \lim_{z \to 0} z^{-\mu}\{g(t)\sigma(z)\} = \lim_{z \to 0} z^{-\mu}\,(g(t)\{g(z)\nu_1\})$$

$$= \lim_{z \to 0} z^{-\mu}g(tz)\,\nu_1 = t^{\mu}\lim_{z \to 0}(tz)^{-\mu}g(tz)\,\nu_1 = t^{\mu}\lim_{z \to 0}(tz)^{-\mu}\sigma(tz) = t^{\mu}\nu_0.$$

Hence μ is the weight w_m of the $\mathbb{C}^*$-action on $\mathcal{N}_0$. As in [4], it then follows that

$$\deg \mathcal{N} = \mu = w_m.$$

Since $\mathscr{E}^{(m)}$ is a vector bundle over $\mathbb{P}^1$ of rank N_m, by a theorem of Birkhoff–Grothendieck, we can write $\mathscr{E}^{(m)}$ as a direct sum

$$\mathscr{E}^{(m)} = \bigoplus_{i=1}^{N_m} \mathscr{O}_{\mathbb{P}^1}(\alpha_i) \tag{2.5}$$

for some integers α_i. Then by the Riemann–Roch theorem for $\mathscr{O}_{\mathbb{P}^1}(\alpha_i)$, we obtain

$$\chi(\mathbb{P}^1, \mathscr{O}_{\mathbb{P}^1}(\alpha_i)) = \deg \mathscr{O}_{\mathbb{P}^1}(\alpha_i) + 1 - g = \alpha_i + 1,$$

since the genus g of $\mathbb{P}^1$ is 0. Hence by (2.4) and (2.5),

$$\chi(\mathbb{P}^1, \mathscr{E}^{(m)}) = \sum_{i=1}^{N_m} \chi(\mathbb{P}^1, \mathscr{O}_{\mathbb{P}^1}(\alpha_i)) = N_m + \sum_{i=1}^{N_m} \alpha_i = N_m + \deg \mathcal{N} = N_m + w_m. \tag{2.6}$$

Since $\mathscr{L}$ is relatively very ample, by $m \gg 1$, we have $R^i\bar{\pi}_*\bar{\mathscr{L}}^{\otimes m} = 0$ for $i > 0$. Hence

$$\chi(\mathbb{P}^1, \mathscr{E}^{(m)}) = \chi(\mathbb{P}^1, \bar{\pi}_*\bar{\mathscr{L}}^{\otimes m}) = \chi(\bar{\mathscr{X}}, \bar{\mathscr{L}}^{\otimes m})$$

$$= \frac{m^{n+1}}{(n+1)!}(\bar{\mathscr{L}}^{n+1}) - \frac{m^{n_i}}{2 \cdot n!}(K_{\bar{\mathscr{X}}} \cdot \bar{\mathscr{L}}^n) + O(m^{n-1}), \tag{2.7}$$

where in the last equality, we used the Riemann–Roch theorem for normal projective varieties [4, Appendix A]. On the other hand, in view of the definition of vol and $\bar{S}$, the Riemann–Roch theorem for the line bundle L shows that

$$
\begin{aligned}
N_m &= \frac{m^n}{n!}\gamma^n c_1(L)^n[X] + \frac{m^{n-1}}{2\cdot(n-1)!}\gamma^{n-1}c_1(X)c_1(L)^{n-1}[X] + O(m^{n-2}) \\
&= \text{vol}\,\frac{m^n}{n!}\left\{1 + \frac{\bar{S}}{2}\cdot\frac{1}{m} + O(\frac{1}{m^2})\right\}.
\end{aligned}
\tag{2.8}
$$

In view of (2.6) together with (2.7) and (2.8), we obtain

$$
\begin{aligned}
w_m &= \chi(\mathbb{P}^1, \mathscr{E}^{(m)}) - N_m \\
&= \frac{m^{n+1}}{(n+1)!}(\bar{\mathscr{L}}^{n+1}) - \frac{m^n}{2\cdot n!}(K_{\bar{\mathscr{X}}}\cdot\bar{\mathscr{L}}^n) - \frac{m^n}{2\cdot n!}2\cdot\gamma^n c_1(L)^n[X] + O(m^{n-1}),
\end{aligned}
$$

where by $(K_{\bar{\mathscr{X}}/\mathbb{P}^1}\cdot\bar{\mathscr{L}}^n) = (K_{\bar{\mathscr{X}}}\cdot\bar{\mathscr{L}}^n) + 2\cdot\gamma^n c_1(L)^n[X]$, we rewrite w_m as

$$
w_m = \frac{m^{n+1}}{(n+1)!}(\bar{\mathscr{L}}^{n+1}) - \frac{m^n}{2\cdot n!}(K_{\bar{\mathscr{X}}/\mathbb{P}^1}\cdot\bar{\mathscr{L}}^n) + O(m^{n-1}).
\tag{2.9}
$$

By (2.8) and (2.9), the terms a_i, b_i, $i = 0, 1$, in the formula (2.3) for $\mathrm{DF}_1(\mathscr{X}, \mathscr{L})$ are given by the following:

$$
a_0 = \frac{\text{vol}}{n!}, \quad a_1 = \frac{\text{vol}\cdot\bar{S}}{2\cdot n!}, \quad b_0 = \frac{(\bar{\mathscr{L}}^{n+1})}{(n+1)!}, \quad b_1 = -\frac{(K_{\bar{\mathscr{X}}/\mathbb{P}^1}\cdot\bar{\mathscr{L}}^n)}{2\cdot n!}.
$$

Hence it follows that

$$
\mathrm{DF}_1(\mathscr{X}, \mathscr{L}) = \frac{a_0 b_1 - a_1 b_0}{a_0^2} = -\frac{1}{2\text{vol}}\left\{(K_{\bar{\mathscr{X}}/\mathbb{P}^1}\cdot\bar{\mathscr{L}}^n) + \frac{\bar{S}}{n+1}(\bar{\mathscr{L}}^{n+1})\right\},
$$

as required. $\square$

2.5   The Relationship Between the Chow Norm and DF_i

Let $(\mathscr{X}, \mathscr{L})$ be a test configuration, of exponent γ, for (X, L). For each positive integer m, we consider the direct image sheaf

$$
\mathscr{E} := \pi_*\mathscr{L}^{\otimes m}, \qquad m = 1, 2, \ldots,
$$

on $\mathbb{A}^1$. Let $m \gg 1$. Then by the affirmative solution of the equivariant Serre conjecture (see Sect. 1.1), $\mathscr{E}$ is $\mathbb{C}^*$-equivariantly isomorphic to $\mathscr{E}_0 \times \mathbb{A}^1$, where

$$\mathscr{E}_z \cong H^0(\mathscr{X}_z, \mathscr{L}_z^{\otimes m}), \qquad z \in \mathbb{A}^1,$$

denotes the fiber of $\mathscr{E}$ over z. Note that $\mathbb{C}^*$ acts on $\mathscr{E}_0 \times \mathbb{A}^1$ by

$$\mathbb{C}^* \times (\mathscr{E}_0 \times \mathbb{A}^1) \ni (t, (e, z)) \mapsto (\sigma_m(t)e, tz) \in \mathscr{E}_0 \times \mathbb{A}^1$$

for an algebraic group homomorphism $\sigma_m : \mathbb{C}^* \to \mathrm{GL}(\mathscr{E}_0)$ naturally induced by the $\mathbb{C}^*$-action on $(\mathscr{X}, \mathscr{L})$. Put $\gamma_m := m\gamma$. Then $X = \mathscr{X}_1 \subset \mathbb{P}(\mathscr{E}_1^*) = \mathbb{P}(V_{\gamma_m}^*)$, where $V_{\gamma_m} = H^0(X, L^{\otimes \gamma_m})$. For the degree d of X in the projective space $\mathbb{P}(V_{\gamma_m}^*)$, we set $W := (\mathrm{Sym}^d V_{\gamma_m}^*)^{\otimes n+1}$, and let $0 \neq \hat{X} \in W$ be the Chow form for X. Let ρ_{γ_m} be the Hermitian structure for V_{γ_m} as in (2.2). By setting $t = \exp s$ for $s \in \mathbb{R}$, we consider the real-valued function f_m on $\mathbb{R}$ defined by

$$f_m(s) := \log \|\sigma_m^{\mathrm{SL}}(t)\hat{X}\|_{\mathrm{CH}(\rho_{\gamma_m})}, \qquad s \in \mathbb{R},$$

where $\| \ \|_{\mathrm{CH}(\rho_{\gamma_m})}$ denotes the Chow norm for W induced by ρ_{γ_m} as in Sect. 1.3. Note that the $\mathbb{C}^*$-equivariant isomorphism $\mathscr{E} \cong \mathscr{E}_0 \times \mathbb{A}^1$ can be chosen in such a way that, by the natural identification $\mathscr{E}_0 \times \{1\} \cong V_{\gamma_m}$, the Hermitian structure ρ_{γ_m} on V_{γ_m} induces a Hermitian structure on $\mathscr{E}_0$ which is preserved by the action of S^1 in $\mathbb{C}^*$. Let a_0 be as in Sect. 2.3. Then by setting $\dot{f}_m(s) = \partial f_m / \partial s$, we obtain (see for instance [51]):

Theorem 2.2 $\displaystyle \lim_{s \to -\infty} \dot{f}_m(s) = (n+1)! \, a_0 \sum_{i=1}^{\infty} \mathrm{DF}_i(\mathscr{X}, \mathscr{L}) m^{n+1-i}.$

Proof For $k = 1, 2, \cdots$, we put $E_k := H^0(\mathscr{X}_0, \mathscr{L}_0^{\otimes km})$ and $V_k := H^0(X, L^{\otimes km\gamma})$. Let d_k be the degree of X in the projective space $\mathbb{P}(V_k^*)$. Then by Mumford [62, Proposition 2.11], the weight μ_k of the $\mathbb{R}_+$-action on $\det E_k$ induced by σ_m^{SL} is

$$\mu_k = -\frac{\lambda_m}{(n+1)!} k^{n+1} + O(k^n), \qquad k \gg 1, \tag{2.10}$$

where λ_m is the Chow weight of $X \subset \mathbb{P}(V_1^*)$, i.e., the weight of the $\mathbb{R}_+$-action on the line $\hat{\mathscr{X}}_0$ in $W_1 := (\mathrm{Sym}^{d_1} E_1^*)^{\otimes n+1}$ via σ_m^{SL}. Note here that $\hat{\mathscr{X}}_0$ is the Chow form for the algebraic cycle $\mathscr{X}_0$ on $\mathbb{P}(E_1^*)(= \mathbb{P}(V_1^*))$. We can write the Chow form $\hat{X} \neq 0$ for the irreducible reduced algebraic cycle $X \subset \mathbb{P}(V_1^*)$ as a sum

$$\hat{X} = \sum_{\alpha=1}^{r} v_\alpha,$$

where $0 \neq v_\alpha \in W_1$ is such that there is an increasing sequence of rational numbers $e_1 < e_2 < \cdots < e_r$ satisfying

$$\sigma_m^{SL}(t) v_\alpha = t^{e_\alpha} v_\alpha, \qquad t \in \mathbb{R}_+.$$

Hence $\hat{\mathscr{X}}_t = \sigma_m^{SL}(t)\hat{\mathscr{X}}_1 = \sigma_m^{SL}(t)\hat{X} = \sum_{\alpha=1}^r t^{e_\alpha} v_\alpha$ for all $t \in \mathbb{C}^*$. Since $\lim_{t\to 0}[\hat{\mathscr{X}}_t] = [\hat{\mathscr{X}}_0]$ in $\mathbb{P}(W_1)$, and since

$$\lim_{t\to 0}[\hat{\mathscr{X}}_t] = \lim_{t\to 0}\left[t^{e_1}\left(v_1 + \sum_{\alpha=2}^r t^{e_\alpha-e_1} v_\alpha\right)\right] = \lim_{t\to 0}\left[v_1 + \sum_{\alpha=2}^r t^{e_\alpha-e_1} v_\alpha\right] = [v_1],$$

we may assume that $\hat{\mathscr{X}}_0 = v_1$. Since λ_m is the weight of the $\mathbb{R}_+$-action on the line $\mathbb{R}\hat{\mathscr{X}}_0$ via σ_m^{SL}, we have

$$e_1 = \lambda_m.$$

It then follows that

$$
\begin{aligned}
f_m(s) &= \log \|\sigma_m^{SL}(t)\hat{X}\|_{CH(\rho_{\tilde{\gamma}})} = \log \|t^{e_1}(v_1 + \sum_{\alpha=2}^r t^{e_\alpha-e_1} v_\alpha)\|_{CH(\rho_{\tilde{\gamma}})} \\
&= \log\left\{\exp(s\lambda_m)\|v_1 + \sum_{\alpha=2}^r t^{e_\alpha-e_1} v_\alpha\|_{CH(\rho_{\tilde{\gamma}})}\right\} \\
&= s\lambda_m + \log\|v_1 + \sum_{\alpha=2}^r t^{e_\alpha-e_1} v_\alpha\|_{CH(\rho_{\tilde{\gamma}})}.
\end{aligned}
$$

Hence by setting $\varphi(s) := \log\|v_1 + \sum_{\alpha=2}^r t^{e_\alpha-e_1} v_\alpha\|_{CH(\rho_{\tilde{\gamma}})}$, we obtain

$$\dot{f}_m(s) = \lambda_m + \dot{\varphi}(s), \qquad s \in \mathbb{R}.$$

Let $s_1 \in \mathbb{R}$ be such that $s_1 \ll -1$. By the mean value theorem, there exists an $\tilde{s}_1 \in \mathbb{R}$ satisfying $-1 + s_1 < \tilde{s}_1 < s_1$ such that

$$\dot{\varphi}(\tilde{s}_1) = \frac{\varphi(s_1) - \varphi(s_1-1)}{s_1 - (s_1-1)} = \varphi(s_1) - \varphi(s_1-1) = \varphi_{|t=e^{s_1}} - \varphi_{|t=e^{s_1-1}}.$$

From the continuity of the Chow norm (cf. [60]), it follows that

$$\lim_{s_1\to-\infty}\dot{\varphi}(\tilde{s}_1) = \lim_{s_1\to-\infty}\{\varphi_{|t=e^{s_1}} - \varphi_{|t=e^{s_1-1}}\} = \varphi_{|t=0} - \varphi_{|t=0} = 0. \qquad (2.11)$$

Since the function $f_m(s)$ is convex ([93]; see also Sect. 1.4), we observe that $\dot{f}_m(s)$ (and hence $\dot{\varphi}(s)$) is non-decreasing in s. Then by (2.11), $\lim_{s\to-\infty} \dot{\varphi}(s) = 0$. Hence

$$\lim_{s\to-\infty} \dot{f}_m(s) = \lambda_m + \lim_{s\to-\infty} \dot{\varphi}(s) = \lambda_m. \tag{2.12}$$

Recall that $N_m = \dim E_1$. Put $N_{km} := \dim E_k$. By $\sigma_m^{\mathrm{SL}}(t) = \{\det \sigma_m(t)\}^{-1/N_m} \sigma_m(t)$, the weight μ_k of the $\mathbb{R}_+$-action on $\det E_k$ (induced by the $\mathbb{R}_+$-action on $E_1 = \mathscr{E}_0$ via the one-parameter group σ_m^{SL}) is expressible as

$$\mu_k = w_{km} - k(w_m/N_m)N_{km}, \tag{2.13}$$

where w_m and w_{km} are the weights of the $\mathbb{C}^*$-action on E_1 and E_k, respectively, induced by the $\mathbb{C}^*$-action via σ_m. Since the natural map of $S^k(E_1)$ to E_k is surjective (cf. [61, 62]) by $m \gg 1$, as far as the $\mathbb{R}_+$-action on E_k is concerned, w_{km} is the weight induced from the $\mathbb{R}_+$-action on $\mathscr{E}_0$ via σ_m, while $-k(w_m/N_m)N_{km}$ is the weight induced from the $\mathbb{R}_+$-action on $\mathscr{E}_0$ via the scalar multiplication by $\{\det \sigma_m(t)\}^{-1/N_m}$. In view of (2.10) and (2.13), for $k \gg 1$, we obtain

$$-\frac{\lambda_m}{(n+1)!}k^{n+1} + O(k^n) = \mu_k = kmN_{km}\left\{\frac{w_{km}}{(km)N_{km}} - \frac{w_m}{mN_m}\right\}$$

$$= kmN_{km}\left\{\sum_{i=0}^{\infty} \mathrm{DF}_i(\mathscr{X}, \mathscr{L})(km)^{-i} - \sum_{i=0}^{\infty} \mathrm{DF}_i(\mathscr{X}, \mathscr{L})m^{-i}\right\}$$

$$= kmN_{km}\left\{\sum_{i=1}^{\infty} \mathrm{DF}_i(\mathscr{X}, \mathscr{L})(km)^{-i} - \sum_{i=1}^{\infty} \mathrm{DF}_i(\mathscr{X}, \mathscr{L})m^{-i}\right\}$$

$$= -kmN_{km}\left\{\sum_{i=1}^{\infty} \mathrm{DF}_i(\mathscr{X}, \mathscr{L})m^{-i} + O(k^{-1})\right\}.$$

Since $N_{km} = a_0(km)^n\{1 + O(k^{-1})\}$, it then follows that

$$\frac{\lambda_m}{(n+1)!} = m^{n+1}a_0 \sum_{i=1}^{\infty} \mathrm{DF}_i(\mathscr{X}, \mathscr{L})m^{-i}.$$

From this together with (2.12), we now conclude that

$$\lim_{s\to-\infty} \dot{f}_m(s) = \lambda_m = (n+1)!a_0 \sum_{i=1}^{\infty} \mathrm{DF}_i(\mathscr{X}, \mathscr{L})m^{n+1-i},$$

as required. $\square$

Problems

2.1 (cf. Li and Xu [37]) For $\mathbb{P}^3(\mathbb{C}) = \{x = (x_0 : x_1 : x_2 : x_3)\}$ and $\mathbb{A}^1 =$ Spec $\mathbb{C}[z]$, we consider the subscheme $\mathscr{X}$ in $\mathbb{P}^3(\mathbb{C}) \times \mathbb{A}^1 = \{(x, z)\}$ defined by the ideal

$$I = (z^2(x_0 + x_3)x_3 - x_2^2, zx_0(x_0 + x_3) - x_1x_2, x_0x_2 - zx_1x_3, x_1^2x_3 - x_0^2(x_0 + x_3)).$$

Let pr_2 be the restriction to $\mathscr{X}$ of the projection $\mathbb{P}^3(\mathbb{C}) \times \mathbb{A}^1 \to \mathbb{A}^1$ to the second factor. Note that $\mathscr{X}_0$ sitting in $\mathbb{P}^3(\mathbb{C})$ is defined by the ideal

$$I_0 = (x_2^2, x_1x_2, x_0x_2, x_1^2x_3 - x_0^2(x_0 + x_3)), \tag{2.14}$$

where $\mathscr{X}_z := \mathrm{pr}_2^{-1}(z)$, $z \in \mathbb{A}^1$, are the scheme-theoretic fibers. Put $\mathscr{L} :=$ $\mathrm{pr}_1^* \, \mathscr{O}_{\mathbb{P}^3}(1)$, where pr_1 is the restriction to $\mathscr{X}$ of the projection $\mathbb{P}^3(\mathbb{C}) \times \mathbb{A}^1 \to \mathbb{P}^3(\mathbb{C})$ to the first factor. For $\mathbb{C}^4 = \mathrm{Spec} \, \mathbb{C}[x_0, x_1, x_2, x_3]$, the group $\mathbb{C}^* = \{t \in \mathbb{C}^*\}$ acts on $\mathbb{C}^4 \times \mathbb{A}^1$ by

$$(x_0, x_1, x_2, x_3, z) \mapsto (x_0, x_1, tx_2, x_3, tz), \qquad t \in \mathbb{C}^*,$$

which naturally induces $\mathbb{C}^*$-actions on $\mathscr{L}$ and $\mathscr{X}$, since $\mathbb{P}^3(\mathbb{C}) = (\mathbb{C}^4 \setminus \{0\})/\mathbb{C}^*$, and since we can identify $\mathscr{O}_{\mathbb{P}^3}(-1)$ with the blow-up of $\mathbb{C}^4$ at the origin. Then it is easy to check that $(\mathscr{X}, \mathscr{L})$ is a test configuration, of exponent $\gamma = 1$, for $(X, L) = (\mathbb{P}^1(\mathbb{C}), \mathscr{O}_{\mathbb{P}^1}(3))$. To see this, show first of all that $\mathscr{X}_1 \cong \mathbb{P}^1(\mathbb{C})$.

2.2 Compute the Donaldson–Futaki invariant $\mathrm{DF}_1(\mathscr{X}, \mathscr{L})$ of the test configuration $(\mathscr{X}, \mathscr{L})$ in Problem 2.1 above.

2.3 In Problem 2.1, consider the normalization $\nu : \tilde{\mathscr{X}} \to \mathscr{X}$ and the pullback $\tilde{\mathscr{L}} := \nu^* \mathscr{L}$. Show that the test configuration $(\tilde{\mathscr{X}}, \tilde{\mathscr{L}})$ induced from $(\mathscr{X}, \mathscr{L})$ is trivial.

Chapter 3
Canonical Kähler Metrics

Abstract In this chapter, we shall introduce various special metrics for compact complex manifolds such as Kähler–Einstein metrics, CSC Kähler metrics, extremal Kähler metrics, Kähler–Ricci solitons and generalized Kähler–Einstein metrics.

- In Sect. 3.1, we give definitions of these special metrics. Here CSC Kähler metrics and extremal Kähler metrics are defined by using the scalar curvature S_ω, while Kähler–Einstein metrics, Kähler–Ricci solitons and generalized Kähler–Einstein metrics are defined by using the Ricci potential f_ω.
- In Sect. 3.2, we shall show that Kähler–Ricci solitons and generalized Kähler–Einstein metrics are Kähler–Einstein analogues in Bakry–Emery geometry by conformal changes via Hamiltonian functions of holomorphic vector fields.

Keywords Kähler–Einstein metrics · CSC Kähler metrics · Extremal Kähler metrics · Kähler–Ricci solitons · Generalized Kähler–Einstein metrics

3.1 Canonical Kähler Metrics on Compact Complex Manifolds

For a Kähler class $\mathscr{K}$ on a compact complex connected manifold X, we choose a Kähler form ω in $\mathscr{K}$. Put $n := \dim X$. By writing ω in terms of the notation in (1.1), we consider the associated Laplacian

$$\Delta_\omega := \sum_{\alpha,\beta} g^{\bar\beta\alpha} \frac{\partial^2}{\partial z_\alpha \partial z_{\bar\beta}}.$$

For the scalar curvature $S_\omega := \mathrm{Tr}_\omega \mathrm{Ric}(\omega)$, we have $\int_X (S_\omega - S_0)\, \omega^n = 0$ by taking the average S_0. Then we have a real-valued smooth function $f_\omega \in C^\infty(X)_{\mathbb{R}}$ such that

$$\Delta_\omega f_\omega = S_\omega - S_0,$$

T. Mabuchi, *Test Configurations, Stabilities and Canonical Kähler Metrics*,
SpringerBriefs in Mathematics, https://doi.org/10.1007/978-981-16-0500-0_3

where $S_0 = (\int_X \omega^n)^{-1} \int_X n \operatorname{Ric}(\omega)\omega^n$ is a constant depending only on the Kähler class and independent of the choice of ω in $\mathscr{K}$. Then the Futaki character [9, 23]

$$\mathscr{F}(y) := \int_X (y f_\omega)\omega^n, \qquad y \in H^0(X, \mathscr{O}(TX)),$$

is independent of the choice of ω in $\mathscr{K}$, and is a Lie algebra character known as an obstruction to the existence of CSC Kähler metrics. Recall that (cf. [8, 9]):

- ω is called *CSC Kähler* if S_ω is constant,
- ω is called *extremal Kähler* if $\operatorname{grad}_\omega^{\mathbb{C}} S_\omega$ is holomorphic.

If ω sits in the Kähler class $c_1(X)$, then up to an additive constant, f_ω coincides with the Ricci potential for ω in Sect. 9.1, and we define (cf. [24, 35, 45]):

- ω is called *Kähler–Einstein* if f_ω is constant,
- ω is called *Kähler–Ricci soliton* if $\operatorname{grad}_\omega^{\mathbb{C}} f_\omega$ is holomorphic,
- ω is called *generalized Kähler–Einstein* if $\operatorname{grad}_\omega^{\mathbb{C}} e^{f_\omega}$ is holomorphic.

However, if the condition $[\omega] = c_1(X)$ is dropped, we don't know much about Kähler–Ricci solitons and generalized Kähler–Einstein metrics in such a broad sense.

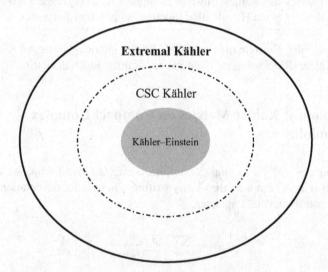

3.2 Conformal Changes of Metrics by Hamiltonian Functions

We now consider the case where $c_1(X) > 0$ and $[\omega] = c_1(X)$. Fix a holomorphic vector field $y \neq 0$ on X. Actually for Kähler–Ricci solitons and generalized Kähler–Einstein metrics, we choose holomorphic vector fields y associated to f_ω and e^{f_ω}, respectively. More generally, a holomorphic vector field y on X written in the form

$$y = \mathrm{grad}_\omega^{\mathbb{C}} \psi,$$

for some $\psi \in C^\infty(X)_{\mathbb{R}}$ is called *Hamiltonian*, where we often impose a normalization condition $\int_X \psi \omega^n = 0$ on ψ. Here ψ is called a *Hamiltonian function* associated to y. In view of the identity $i_y \omega = \bar{\partial}(\psi/2\pi)$, the real vector field $y_{\mathbb{R}} := y + \bar{y}$ on X satisfies $i_{y_{\mathbb{R}}} \omega = d(\psi/2\pi)$. Hence we have a *moment map* for y,

$$\psi : X \to \mathbb{R},$$

where by the normalization condition, both $a := \min_X \psi$ and $b := \max_X \psi$ are constants depending only on y, and independent of the choice of ω in the class $c_1(X)$.

Next we consider a smooth function $\zeta = \zeta(s) : [a, b] \to \mathbb{R}$, where a normalization condition on the Hamiltonian function ψ is not necessarily imposed. Then we put $\zeta(s) = s$ for Kähler–Ricci solitons, whereas we put $\zeta(s) = \log s$ for generalized Kähler–Einstein metrics. In view of the counterpart (cf. [24, 47]) in complex geometry of the Bakry–Emery theory [2, 67, 84], by setting $\tilde{\omega} := \exp\{\zeta(\psi)/n\} \omega$, we can write the corresponding volume form as $\tilde{\omega}^n = e^{\zeta(\psi)} \omega^n$, so that we define

$$\mathrm{Ric}(\tilde{\omega}) := -dd^c \log \tilde{\omega}^n = \mathrm{Ric}(\omega) - dd^c \zeta(\psi) = \omega + dd^c \{f_\omega - \zeta(\psi)\}.$$

Then the *Einstein condition* $\mathrm{Ric}(\tilde{\omega}) = \omega$ is the constancy of $f_\omega - \zeta(\psi)$. Since the Ricci potential f_ω for ω is unique up to an additive constant, for a suitable choice of an additive constant, we can express the Einstein condition as

$$f_\omega = \zeta(\psi) \quad \text{for some } \psi.$$

Hence ω satisfying the Einstein condition $\mathrm{Ric}(\tilde{\omega}) = \omega$ is a Kähler–Ricci soliton or a generalized Kähler–Einstein metric, if $\zeta(s) = s$ or $\zeta(s) = \log s$, respectively.

Problems

3.1 Let $y \in \mathfrak{k} := \mathrm{Lie}(K)$ be a real vector field on a compact connected smooth manifold X with an effective action of a compact real Lie group K. By choosing a K-invariant real symplectic form ω on X, assume that some $f \in C^\infty(X)_{\mathbb{R}}$ satisfies

$$df = i_y \omega.$$

Show that the value $\max_X f - \min_X f$ is independent of the choice of ω as long as ω defines the same de Rham cohomology class $[\omega]$.

3.2 For $\tilde{\omega} = \exp\{\zeta(\psi)/n\}\,\omega$ as above, let $\Delta_{\tilde{\omega}} := \Delta_\omega - \sqrt{-1}\dot{\zeta}(\psi)\,\bar{y}$ be the associated Laplacian, where $\dot{\zeta}(s) := d\zeta(s)/ds$. Then for all $\xi, \eta \in C^\infty(X)_\mathbb{R}$, show that $-\int_X (\bar{\partial}\xi, \bar{\partial}\eta)_\omega\,\tilde{\omega}^n = \int_X (\Delta_{\tilde{\omega}}\xi)\eta\tilde{\omega}^n$, where $(\ ,\ )_\omega$ denotes the pointwise Hermitian pairing by ω for 1-forms on X.

3.3 For a constant scalar curvature Kähler metric ω, let y be a Hamiltonian holomorphic vector field, i.e., $y = \operatorname{grad}_\omega^\mathbb{C} \psi$ for some $\psi \in C^\infty(X)_\mathbb{R}$. Show that $\mathscr{F}(y) = 0$ for the Futaki character $\mathscr{F}$.

Chapter 4
Norms for Test Configurations

Abstract In this chapter, we introduce the concept of norms for test configurations. Then by using such norms, we can define the Donaldson–Futaki invariant for sequences of test configurations.

- In Sect. 4.1, we define norms of test configurations of a fixed exponent.
- In Sect. 4.2, given a test configuration $(\mathscr{X}, \mathscr{L})$, we obtain an asymptotic ℓ^1-norm from the sequence of test configurations $(\mathscr{X}, \mathscr{L}^{\otimes m})$, $m = 1, 2, \cdots$.
- In Sect. 4.3, we consider the relative version of norms of test configurations.
- In Sect. 4.4, we define twisted Kodaira embeddings.
- Finally, in Sect. 4.5, we define the Donaldson–Futaki invariant for sequences of test configurations.

Keywords Norms for test configurations · Twisted Kodaira embeddings · The Donaldson–Futaki invariant for sequences of test configurations

4.1 Norms for Test Configurations of a Fixed Exponent

For a polarized algebraic manifold (X, L), we fix a Hermitian metric h for L such that $\omega := \mathrm{Ric}(h)$ is Kähler. Let $\mu_\sigma = (\mathscr{X}_\sigma, \mathscr{L}_\sigma)$ be a test configuration, of exponent γ, associated to a special one-parameter group

$$\sigma \; : \; \mathbb{C}^* \to \mathrm{GL}(V_\gamma).$$

Here for the subgroup S^1 of $\mathbb{C}^*$, $\sigma(S^1)$ acts isometrically on the vector space $V_\gamma := H^0(X, L^{\otimes \gamma})$ endowed with the Hermitian metric ρ_γ in (2.2). Put $N_\gamma := \dim V_\gamma$. As in Sect. 1.1, for σ, we consider its special linearization σ^{SL} such that, when restricted to $\mathbb{R}_+$, we obtain a Lie group homomorphism

$$\sigma^{\mathrm{SL}} \; : \; \mathbb{R}_+ \to \mathrm{SL}(V_\gamma).$$

T. Mabuchi, *Test Configurations, Stabilities and Canonical Kähler Metrics*, SpringerBriefs in Mathematics, https://doi.org/10.1007/978-981-16-0500-0_4

Let $u \in \mathfrak{sl}(V_\gamma)$ be the fundamental generator (cf. Sect. 1.1) for σ^{SL}. Then for a suitable choice of an orthonormal basis for V_γ, u is written in the form

$$
u = \begin{pmatrix} -\beta_1 & & & 0 \\ & -\beta_2 & & \\ & & \cdots & \\ 0 & & & -\beta_{N_\gamma} \end{pmatrix},
$$

where $-\beta_i$, $i = 1, 2, \ldots, N_\gamma$, are the weights of the $\mathbb{R}_+$-action on V_γ via σ^{SL}, so that $\sigma(e^s) = \exp(su)$ holds for all $s \in \mathbb{R}$. We now put

$$
|u|_1 := \gamma^{-1} N_\gamma^{-1} \left(\sum_{i=1}^{N_\gamma} |\beta_i| \right), \quad |u|_\infty := \gamma^{-1} \max_{1 \le i \le N_\gamma} |\beta_i|. \tag{4.1}
$$

Then $|u|_1 = 0$ if and only if $u = 0$, i.e., $(\mathscr{X}_\sigma, \mathscr{L}_\sigma)$ is trivial. The same thing is true also for $|u|_\infty$. By abuse of terminology, we define the following:

Definition 4.1 The ℓ^1-norm $\|\mu_\sigma\|_1$ of the test configuration $\mu_\sigma = (\mathscr{X}_\sigma, \mathscr{L}_\sigma)$ is $|u|_1$, while the ℓ^∞-norm $\|\mu_\sigma\|_\infty$ of the test configuration $\mu_\sigma = (\mathscr{X}_\sigma, \mathscr{L}_\sigma)$ is $|u|_\infty$.

4.2   The Asymptotic ℓ^1-norm of a Test Configuration

Let $\mu = (\mathscr{X}, \mathscr{L})$ be a test configuration, of exponent γ, for a polarized algebraic manifold (X, L). As in Sect. 2.3, the algebraic torus $\mathbb{C}^*$ acts on $H^0(\mathscr{X}_0, \mathscr{L}_0^{\otimes m})$. Then as in Sect. 2.5, this action is given by a representation

$$
\sigma_m : \mathbb{C}^* \to \mathrm{GL}(H^0(\mathscr{X}_0, \mathscr{L}_0^{\otimes m})), \qquad m \gg 1,
$$

associated to the test configuration $(\mathscr{X}, \mathscr{L}^{\otimes m})$. By Sect. 1.1, we have its special linearization σ_m^{SL}, and when restricted to $\mathbb{R}_+$, we obtain a Lie group homomorphism

$$
\sigma_m^{\mathrm{SL}} : \mathbb{R}_+ \to \mathrm{SL}(H^0(\mathscr{X}_0, \mathscr{L}_0^{\otimes m})).
$$

Put $\gamma_m := m\gamma$ and $N_m := \dim H^0(\mathscr{X}_0, \mathscr{L}_0^{\otimes m})$. Let $-\beta_i$, $i = 1, 2, \cdots, N_m$, be the weights of the $\mathbb{R}_+$-action on $H^0(\mathscr{X}_0, \mathscr{L}_0^{\otimes m})$. Then the fundamental generator u_m for the special linearization σ_m^{SL} is written as

$$
u_m = \begin{pmatrix} -\beta_1 & & & 0 \\ & -\beta_2 & & \\ & & \cdots & \\ 0 & & & -\beta_{N_m} \end{pmatrix}
$$

by choosing a suitable basis for $H^0(\mathscr{X}_0, \mathscr{L}_0^{\otimes m})$. Then a result of Hisamoto [30] shows that the following limit exists as an *asymptotic ℓ^1-norm* $\|\mu\|_{asymp}$ of μ:

$$\|\mu\|_{asymp} := \lim_{m \to \infty} |u_m|_1 = \lim_{m \to \infty} \left(\gamma_m^{-1} N_m^{-1} \sum_{i=1}^{N_m} |\beta_i| \right).$$

Remark 4.1 By definition, $\|(\mathscr{X}, \mathscr{L}^{\otimes k})\|_{asymp} = \|(\mathscr{X}, \mathscr{L})\|_{asymp}$ for all positive integers k. It is known that, as long as $\mathscr{X}$ is a normal complex variety, $\|\mu\|_{asymp} = 0$ if and only if the test configuration μ is trivial ([4]; see also [18, 30]).

4.3 Relative Norms for Test Configurations

As a reference for this section, see [76] (cf. also [54]). For a polarized algebraic manifold (X, L), we consider a possibly trivial algebraic torus T sitting in the identity component $\mathrm{Aut}^0(X)$ of $\mathrm{Aut}(X)$. Replacing L by its suitable power, we may assume that the T-action on X lifts to an action on L inducing linear maps between fibers. To study extremal Kähler metrics on X, we consider $T_\gamma^{\perp}$ below in place of $G_\gamma := \mathrm{SL}(V_\gamma)$. Now for $\mathfrak{g}_\gamma := \mathrm{Lie}(G_\gamma) = \mathfrak{sl}(V_\gamma)$, we define a bilinear form $\langle \, , \, \rangle_\gamma$ by

$$\langle u, v \rangle_\gamma := \gamma^{-n-2} \mathrm{Tr}(uv), \qquad u, v \in \mathfrak{g}_\gamma. \tag{4.2}$$

Let $\mathfrak{t}_\gamma$ be the Lie algebra $\mathfrak{t} := \mathrm{Lie}(T)$ viewed as a Lie subalgebra of $\mathfrak{g}_\gamma$. Let $\mathfrak{z}_\gamma := \{u \in \mathfrak{g}_\gamma ; [u, \mathfrak{t}_\gamma] = 0\}$ be its centralizer in $\mathfrak{g}_\gamma$. Then for $\mathfrak{t}_\gamma$ and $\mathfrak{z}_\gamma$, the corresponding connected linear algebraic subgroups in G_γ will be denoted by T_γ and Z_γ, respectively. Let $\mathfrak{t}_\gamma^{\perp}$ be the orthogonal complement of $\mathfrak{t}_\gamma$ in $\mathfrak{z}_\gamma$ defined by

$$\mathfrak{t}_\gamma^{\perp} := \{u \in \mathfrak{z}_\gamma ; \langle u, \mathfrak{t}_\gamma \rangle_\gamma = 0\}.$$

Note that, if T is trivial, then $\mathfrak{t}_\gamma^{\perp} = \mathfrak{z}_\gamma = \mathfrak{g}_\gamma$. To see another expression of $\mathfrak{t}_\gamma^{\perp}$, by considering the infinitesimal $\mathfrak{t}_\gamma$-action on V_γ, we write V_γ as a direct sum

$$V_\gamma = \bigoplus_{i=1}^{n_\gamma} V_{\gamma,i},$$

where $V_{\gamma,i} = \{v \in V_\gamma ; \theta v = \chi_{\gamma,i}(\theta)v \text{ for all } \theta \in \mathfrak{t}_\gamma\}$ for distinct characters $\chi_{\gamma,i} \in \mathfrak{t}_\gamma^*, i = 1, \cdots, n_\gamma$. By choosing a basis for each $V_{\gamma,i}$, we identify G_γ with $\mathrm{SL}(N_\gamma, \mathbb{C})$. Put

$$S_\gamma := \prod_{i=1}^{n_\gamma} \mathrm{SL}(V_{\gamma,i}) \subset \mathrm{SL}(N_\gamma, \mathbb{C}).$$

The centralizer H_γ of S_γ in G_γ consists of all diagonal matrices in G_γ which act on each $V_{\gamma,i}$, $i = 1, \ldots, n_\gamma$, by constant scalar multiplications. By setting $\mathfrak{h}_\gamma :=$ $\mathrm{Lie}(H_\gamma)$, we consider the orthogonal complement of $\mathfrak{t}_\gamma$ in $\mathfrak{h}_\gamma$,

$$\mathfrak{t}'_\gamma := \{ u \in \mathfrak{h}_\gamma \, ; \, \langle u, \mathfrak{t}_\gamma \rangle_\gamma = 0 \},$$

and the associated algebraic torus in H_γ will be denoted by T'_γ. Then

$$T_\gamma^\perp := T'_\gamma \cdot S_\gamma$$

is a reductive algebraic subgroup of G_γ whose Lie algebra is $\mathfrak{t}_\gamma^\perp$. Let T_c be the maximal compact subgroup of T. Let $\omega = \mathrm{Ric}(h)$ be a T_c-invariant Kähler form on X, so that we can choose h as a T_c-invariant Hermitian metric for L.

Definition 4.2 For T above, we denote by $\mathrm{1PS}(T_\gamma^\perp)$ the set of all nontrivial special one-parameter groups $\sigma : \mathbb{C}^* \to \mathrm{SL}(V_\gamma)$ satisfying $\sigma(\mathbb{C}^*) \subset T_\gamma^\perp$.

Remark 4.2 If the algebraic torus T is trivial, then $\mathrm{1PS}(T_\gamma^\perp)$ is nothing but the set $\mathrm{1PS}(G_\gamma)$ of all nontrivial special one-parameter groups $\sigma : \mathbb{C}^* \to G_\gamma := \mathrm{SL}(V_\gamma)$.

For each $\sigma \in \mathrm{1PS}(T_\gamma^\perp)$, let $u \in \mathfrak{t}_\gamma^\perp$ be its fundamental generator. By Sect. 2.2, we have the test configuration

$$\mu_\sigma = (\mathscr{X}_\sigma, \mathscr{L}_\sigma)$$

associated to the special one-parameter group σ. As in Sect. 4.2, for each positive integer m, the one-parameter group σ induces

$$u_m \in \mathfrak{sl}(H^0((\mathscr{X}_\sigma)_0, (\mathscr{L}_\sigma^{\otimes m})_0)).$$

As in Sect. 2.5, $H^0((\mathscr{X}_\sigma)_0, (\mathscr{L}_\sigma^{\otimes m})_0)$ is viewed as $H^0((\mathscr{X}_\sigma)_1, (\mathscr{L}_\sigma^{\otimes m})_1) = V_{\gamma_m}$ with $\gamma_m := m\gamma$, so that we regard u_m as an element of $\mathfrak{sl}(V_{\gamma_m})$ with real eigenvalues. Put $\mathfrak{t}_\mathbb{R} := \sqrt{-1}\, \mathfrak{t}_c$, where $\mathfrak{t}_c$ is the maximal compact real Lie subalgebra of $\mathfrak{t}$. Then for each $v \in \mathfrak{t}_\mathbb{R}$, the element in $\mathfrak{sl}(V_{\gamma_m})$ induced by v will be denoted as v_m. Since u commutes with $\mathfrak{t}_\gamma$, we see that u_m and v_m are simultaneously diagonalized. Hence by choosing a suitable basis for V_{γ_m}, we can write

$$u_m = \begin{pmatrix} -\beta_1 & & & 0 \\ & -\beta_2 & & \\ & & \ddots & \\ 0 & & & -\beta_{N_m} \end{pmatrix}, \quad v_m = \begin{pmatrix} \alpha_1 & & & 0 \\ & \alpha_2 & & \\ & & \ddots & \\ 0 & & & \alpha_{N_m} \end{pmatrix}, \qquad (4.3)$$

where α_i, β_j, $i, j \in \{1, 2, \cdots N_m\}$, are real constants. In [30], replacing the fundamental generator u_m by $u_m + v_m$, we obtain $\|\mu_\sigma\|_{asymp}^v$ by setting

$$\|\mu_\sigma\|_{asymp}^v := \lim_{m\to\infty} |u_m + v_m|_1 = \lim_{m\to\infty} \left(\gamma_m^{-1} N_m^{-1} \sum_{i=1}^{N_m} |\alpha_i - \beta_i| \right).$$

Then a result in Hisamoto [31] shows that the *asymptotic ℓ^1-norm* $\|\mu_\sigma\|_{asymp}^T$ for μ_σ *relative to T* is defined by

$$\|\mu_\sigma\|_{asymp}^T := \inf_{v\in t_\mathbb{R}} \|\mu_\sigma\|_{asymp}^v > 0.$$

Remark 4.3 If T is trivial, then $\|\mu\|_{asymp}^T$ coincides with $\|\mu\|_{asymp}$ in Sect. 4.2.

Remark 4.4 For $r := \dim t_\mathbb{R}$, we choose elements w_i, $i = 1, \ldots, r$, in $t_\mathbb{R}$ such that the corresponding elements $w_{i,m}$, $i = 1, \ldots, r$, in t_{γ_m} satisfy

$$\langle w_{i,m}, w_{j,m} \rangle_{\gamma_m} = \delta_{ij}, \qquad i, j \in \{1, 2, \cdots, r\},$$

where δ_{ij} is Kronecker's delta, and $\langle \cdot, \rangle_{\gamma_m}$ is the bilinear form on g_{γ_m} as in (4.2). Then for u_m in (4.3), we define $u_m' \in \mathfrak{sl}(V_{\gamma_m})$ with real eigenvalues by

$$u_m' := u_m - \sum_{i=1}^{r} \langle u_m, w_{i,m} \rangle_{\gamma_m} w_{i,m}.$$

In place of the above definition of $\|\mu_\sigma\|_{asymp}^T$, it is possible to use the formula

$$\|\mu_\sigma\|_{asymp}^T := \lim_{m\to\infty} |u_m'|_1.$$

This new definition and the original definition for $\|\mu_\sigma\|_{asymp}^T$ are known to give equivalent norms (see [31] for more details).

4.4 The Twisted Kodaira Embedding

For a polarized algebraic manifold (X, L), let y be the extremal vector field for (X, L) as in Sect. 9.2. By the notation in Sect. 4.3, we choose an orthonormal basis

$$\{v_{i,\alpha} \, ; \, i = 1, 2, \cdots, n_\gamma, \, \alpha = 1, 2, \cdots, q_i\}$$

for (V_γ, ρ_γ) such that each $\{v_{i,\alpha}\,;\, \alpha = 1, 2, \cdots, q_i\}$, $i = 1, 2, \cdots, n_\gamma$, is a basis for $V_{\gamma,i}$. Let y_γ be the element in $\mathfrak{t}_\gamma$ corresponding to y in $\mathfrak{t}$. Put

$$v_{i,\alpha}^\# := \{1 - \gamma^{-2}\chi_{\gamma,i}(\sqrt{-1}y_\gamma)\}^{1/2}v_{i,\alpha}.$$

Renumber the basis $\{v_{i,\alpha}^\#\,;\, i = 1, 2, \cdots, n_\gamma,\ \alpha = 1, 2, \cdots, q_i\}$ as $\{v_1^\#, v_2^\#, \cdots, v_{N_\gamma}^\#\}$. This renumbered basis $\{v_1^\#, v_2^\#, \cdots, v_{N_\gamma}^\#\}$ is called an *admissible basis* for V_γ. Then the *twisted Kodaira embedding* $\Phi_\gamma^\#$ for $L^{\otimes\gamma}$ is

$$\Phi_\gamma^\# : X \to \mathbb{P}(V_\gamma^*), \qquad x \mapsto (v_1^\#(x) : v_2^\#(x) : \cdots : v_{N_\gamma}^\#(x)),$$

where $\mathbb{P}(V_\gamma^*)$ is viewed as the complex projective space $\mathbb{P}^{N_\gamma-1}(\mathbb{C})$ in terms of the admissible basis. Here by $\gamma^{-2}(\chi_{\gamma,i})_*(\sqrt{-1}y_\gamma) = O(\gamma^{-1}) \in \mathbb{R}$ (cf. [50, 54]), we have

$$v_{i,\alpha}^\# = \{1 + O(\gamma^{-1})\}\,v_{i,\alpha},$$

and hence $\Phi_\gamma^\#$ above is well-defined for $\gamma \gg 1$. If the extremal vector field y is trivial, then $\Phi_\gamma^\#$ coincides with the original Kodaira embedding

$$\Phi_\gamma : X \to \mathbb{P}(V_\gamma^*), \qquad x \mapsto (v_1(x) : v_2(x) : \cdots : v_{N_\gamma}(x)).$$

4.5   The Donaldson–Futaki Invariant for Sequences

In this section, we give an example where the norms $|u|_1$ and $|u|_\infty$ in (4.1) are needed. For a polarized algebraic manifold (X, L), we choose a possibly trivial algebraic torus T in $\mathrm{Aut}^0(X)$. Let $\mathcal{M}_T$ denote the set of all sequences $\{\mu_j\}$ of test configurations $\mu_j = (\mathcal{X}_j, \mathcal{L}_j)$, $j = 1, 2, \cdots$, for (X, L) such that the exponent γ_j of μ_j satisfies

$$\gamma_j \to +\infty, \quad \text{as } j \to \infty,$$

where each μ_j is a test configuration associated to some $\sigma_j \in 1\mathrm{PS}(T_{\gamma_j}^\perp)$. In the special case where T is trivial, $T_{\gamma_j}^\perp$ is nothing but G_{γ_j} in terms of the notation in Remark 4.2, and $\mathcal{M}_T$ will be written simply as $\mathcal{M}$. In this section, we consider $\mathcal{M}_T$ for a general T. Let $0 \neq u_j \in \mathfrak{t}_{\gamma_j}$ be the fundamental generator for σ_j. By setting

$t := \exp(s/|u_j|_\infty)$, $s \in \mathbb{R}$, we define a real-valued function $f_j(s)$ on $\mathbb{R}$ by

$$f_j(s) := \frac{|u_j|_\infty}{|u_j|_1} \gamma_j^{-n} \log \|\sigma_j(t) \cdot \hat{X}_j\|_{\mathrm{CH}(\rho_{\gamma_j})}$$

$$= \frac{|u_j|_\infty}{|u_j|_1} \gamma_j^{-n} \log \| \exp(su_j/|u_j|_\infty)\hat{X}_j\|_{\mathrm{CH}(\rho_{\gamma_j})}, \tag{4.4}$$

where for the twisted Kodaira embedding $\Phi_{\gamma_j}^{\#} : X \hookrightarrow \mathbb{P}(V_{\gamma_j}^*)$ for $L^{\otimes \gamma_j}$ on $X = (\mathscr{X}_j)_1$, we consider the Chow form $\hat{X}_j$ for the irreducible reduced algebraic cycle $X_j := \Phi_{\gamma_j}^{\#}(X)$ on $\mathbb{P}(V_{\gamma_j}^*)$. We now claim the following:

Claim $\dot{f}_j(0) \leq C$ for some real constant $C > 0$ independent of j.

Assuming this claim, we now define an invariant $F_1(\{\mu_j\}) \in \mathbb{R} \cup \{-\infty\}$, called *the Donaldson–Futaki invariant for* $\{\mu_j\}$, as follows: For each $s \leq 0$, we have

$$\varliminf_{j\to\infty} \dot{f}_j(s) \leq \varliminf_{j\to\infty} \dot{f}_j(0) \leq C,$$

since $\varliminf_{j\to\infty} \dot{f}_j(s)$ is a non-decreasing function of s by the convexity of $f_j(s)$. Then by letting $s \to -\infty$, we can define the following (cf. [52]):

$$F_1(\{\mu_j\}) := \lim_{s\to-\infty} \varliminf_{j\to\infty} \dot{f}_j(s) \leq C. \tag{4.5}$$

Proof of Claim Since σ_j is a special one-parameter group, by choosing an admissible basis $\{v_1^{\#}, \cdots, v_{N_j}^{\#}\}$ and the associated orthonormal basis $\{v_1, \cdots, v_{N_j}\}$ for the Hermitian vector space $(V_{\gamma_j}, \rho_{\gamma_j})$, we have integers β_k such that for all $t \in \mathbb{C}^*$,

$$\sigma_j(t) \cdot v_k = t^{-\beta_k} v_k, \qquad k = 1, \cdots, N_j,$$

where $\beta_1 + \cdots + \beta_{N_j} = 0$. Let $B_j(\omega) := (n!/\gamma_j^n) \sum_{k=1}^{N_j} |v_k|_h^2$ be the γ_j-th asymptotic Bergman kernel, and we further put

$$B_j^{\#}(\omega) := (n!/\gamma_j^n) \sum_{k=1}^{N_j} |v_k^{\#}|_h^2.$$

Then by the theorem of Tian–Yau–Zelditch [77, 92] and Lu [39],

$$B_j(\omega) = 1 + \frac{1}{2} S_\omega \gamma_j^{-1} + O(\gamma_j^{-2}) = 1 + O(\gamma_j^{-1}), \tag{4.6}$$

so that we can write $B_j^\#(\omega)$ as

$$B_j^\#(\omega) \;=\; \{1 + O(\gamma_j^{-1})\}^2 B_j(\omega) \;=\; 1 + O(\gamma_j^{-1}). \tag{4.7}$$

Then by taking dd^c log of both sides of this equality, we obtain

$$\omega_{\mathrm{FS}} - \gamma_j \omega \;=\; O(\gamma_j^{-1}), \tag{4.8}$$

where $\omega_{\mathrm{FS}} = dd^c \log \sum_{k=1}^{N_j} |v_k^\#|^2$. Apply [48, Remark 4.6] to (1.8). Then by (4.8),

$$
\begin{aligned}
\dot{f}_j(0) \;&=\; \frac{|u_j|_\infty}{|u_j|_1} \gamma_j^{-n} \frac{d}{ds}_{|s=0} \log \|\sigma_j(\exp(s/|u_j|_\infty)\hat{X}_j\|_{\mathrm{CH}(\rho_{\gamma_j})} \\
&=\; \frac{1}{|u_j|_1} \gamma_j^{-n}(n+1) \int_X \frac{\beta_1 |v_1^\#|_h^2 + \cdots + \beta_{N_j} |v_{N_j}^\#|_h^2}{|v_1^\#|_h^2 + \cdots + |v_{N_j}^\#|_h^2} \omega_{\mathrm{FS}}^n \\
&=\; \frac{(n+1)!}{|u_j|_1} \gamma_j^{-n} \int_X \frac{\sum_{k=1}^{N_j} \beta_k |v_k|_h^2 \{1 + O(\gamma_j^{-1})\}}{B_j^\#(\omega)} \{\omega + O(\gamma_j^{-2})\}^n.
\end{aligned}
\tag{4.9}
$$

Since $\{v_1, \cdots, v_{N_j}\}$ is an orthonormal basis for $(V_{\gamma_j}, \rho_{\gamma_j})$, we have the identity $\int_X \sum_{k=1}^{N_j} \beta_k |v_k|_h^2 \omega^n = \sum_{k=1}^{N_j} \beta_k = 0$, and hence by (4.7) and (4.9) together with (4.1),

$$
\begin{aligned}
\dot{f}_j(0) \;&=\; \frac{(n+1)!}{|u_j|_1} \gamma_j^{-n} \int_X O(\gamma_j^{-1}) \sum_{k=1}^{N_j} \left(|\beta_k| \cdot |v_k|_h^2 \right) \omega^n \\
&=\; \frac{(n+1)!}{|u_j|_1} O(\gamma_j^{-n-1}) \sum_{k=1}^{N_j} |\beta_k| \;=\; O(1). \qquad\qquad \square
\end{aligned}
$$

Remark 4.5 In (4.4) above, if we replace u_j by its constant scalar multiple $c_j u_j$, the function $f_j(s)$ is still the same. Hence given $\mu_j = (\mathcal{X}_{\sigma_j}, \mathcal{L}_{\sigma_j})$, $j = 1, 2, \cdots$, we can define $F_1(\{\mu_j\})$ more generally even when each σ_j has rational weights, because by constant scalar multiplication, a rational matrix is changed to an integral matrix. Hence even in the case where each σ_j is a special one-parameter group from $\mathbb{C}^*$ to $\mathrm{GL}(V_{\gamma_j})$, we can still define $F_1(\{\mu_j\})$ by setting

$$f_j(s) \;:=\; \frac{|u_j|_\infty}{|u_j|_1} \gamma_j^{-n} \log \|\sigma_j^{\mathrm{SL}}(t) \cdot \hat{X}_j\|_{\mathrm{CH}(\rho_{\gamma_j})}, \tag{4.10}$$

where u_j is the fundamental generator for σ_j^{SL}. In this case, σ_j is not necessarily an algebraic group homomorphism from $\mathbb{C}^*$ to $SL(V_{\gamma_j})$, but its lift

$$\tilde{\sigma}_j : \tilde{\mathbb{C}}^* \to SL(V_{\gamma_j})$$

to a suitable covering $\tilde{\mathbb{C}}^*$ of the algebraic torus $\mathbb{C}^*$ is an algebraic group homomorphism. It is then easy to see that $F_1(\{\mu_j\})$ thus obtained from the function $f_j(s)$ in (4.10) is nothing but $F_1(\{\mu_{\tilde{\sigma}_j}\})$.

Remark 4.6 In the definition of $F_1(\{\mu_j\})$, all test configurations $\mu_j = (\mathscr{X}_{\sigma_j}, \mathscr{L}_{\sigma_j})$ are assumed to be nontrivial, i.e., nontriviality of the one-parameter groups $\sigma_j :$ $\mathbb{C}^* \to T_{\gamma_j}^\perp$ is assumed. However, we can drop this assumption as follows: If μ_j is a trivial test configuration, we choose $f_j(s)$ as a constant function by setting

$$f_j(s) := \|\hat{X}_j\|_{CH(\rho_{\gamma_j})}, \qquad s \in \mathbb{R}.$$

Then we can define $F_1(\{\mu_j\})$ by the formula (4.5) even if the condition of nontriviality of μ_j is not assumed.

Problems

4.1 For the test configuration $\mu = (\mathscr{X}, \mathscr{L})$ for $(X, L) = (\mathbb{P}^1(\mathbb{C}), \mathscr{O}_{\mathbb{P}^1}(3))$ in Problem 2.1, compute the asymptotic ℓ^1-norm $\|\mu\|_{asymp}$.

4.2 Let $(X, L) = (\mathbb{P}^1(\mathbb{C}), \mathscr{O}_{\mathbb{P}^1}(1))$ for $\mathbb{P}^1(\mathbb{C}) = \{(x_0 : x_1)\}$. Consider the product test configuration $\mu = (\mathscr{X}, \mathscr{L})$ of exponent 1 for (X, L), where by the representation

$$\mathbb{C}^* \ni t \mapsto \begin{pmatrix} t & 0 \\ 0 & t^{-1} \end{pmatrix} \in SL(H^0(X, L)),$$

the algebraic torus $\mathbb{C}^*$ acts on $H^0(X, L)$ in terms of the basis $\{x_0, x_1\}$ for $H^0(X, L)$. Compute the asymptotic ℓ^1-norm $\|\mu\|_{asymp}$ for μ.

4.3 For $\{\mu_j\}_{j=1,2,\cdots} \in \mathscr{M}_T$, let $\{\mu_{j_k}\}_{k=1,2,\cdots}$ be its subsequence. Show that

$$F_1(\{\mu_j\}) \le F_1(\{\mu_{j_k}\}).$$

Chapter 5
Stabilities for Polarized Algebraic Manifolds

Abstract In this chapter, several stability concepts will be introduced from the viewpoints of the existence problem of canonical Kähler metrics.

- Typical examples of such stabilities are the Chow stability and the Hilbert stability. We shall first study these classical stability concepts by showing that they are asymptotically equivalent.
- Secondly, various kinds of K-stability will be discussed to study the existence problem of Kähler-Einstein metrics or more generally CSC Kähler metrics.
- Thirdly, we introduce relative versions of stability concepts, which play a crucial role in the study of extremal Kähler metrics.
- Finally, various relationships among the stability concepts will be discussed.

Keywords The Chow stability · The Hilbert stability · K-stability · Relative stability

5.1 The Chow Stability

Let $X \subset \mathbb{P}(V^*)$ be an n-dimensional projective subvariety, where V is a finite dimensional complex vector space. As in Sect. 1.4, let

$$0 \neq \hat{X} \in W$$

be the Chow form for X, where $W := (\mathrm{Sym}^d V^*)^{\otimes n+1}$ and $d := \deg_{\mathbb{P}(V^*)} X$. For a reductive algebraic subgroup G of $\mathrm{SL}(V)$, consider the induced G-action on $\mathbb{P}(V^*)$.

Definition 5.1

(1) $X \subset \mathbb{P}(V^*)$ is called *Chow polystable* if the orbit $G \cdot \hat{X}$ is closed in the vector space W.
(2) Let $X \subset \mathbb{P}(V^*)$ be Chow polystable. Then $X \subset \mathbb{P}(V^*)$ is called *Chow stable* if in addition the isotropy subgroup of G at $\hat{X}$ is finite.

T. Mabuchi, *Test Configurations, Stabilities and Canonical Kähler Metrics*,
SpringerBriefs in Mathematics, https://doi.org/10.1007/978-981-16-0500-0_5

From now on in this section, consider the case where $V = V_\gamma := H^0(X, L^{\otimes\gamma})$ for a polarized algebraic manifold (X, L). Then we consider the Kodaira embedding

$$X \subset \mathbb{P}(V_\gamma^*), \qquad x \mapsto \text{hyperplane } \{v \in V_\gamma ; \; v(x) = 0\} \text{ in } V_\gamma,$$

associated to the complete linear system $|L^{\otimes\gamma}|$, where this embedding is just an abstract one without assuming any orthonormality of the basis for the embedding. Note that the ordinary Kodaira embedding and the twisted Kodaira embedding coincide as an abstract Kodaira embedding.

Assume that $G = \mathrm{SL}(V)$. For a Kähler form ω in $c_1(L)$, choose a Hermitian metric h for L such that $\mathrm{Ric}(h) = \omega$. Hence, in terms of the Hermitian structure ρ_γ for $V = V_\gamma$ as in (2.2), we can talk about a special one-parameter group. For each $v \in V$, we consider the nonnegative function $|v|_h$ on X defined by $|v|_h^2 = (v, v)_h$.

Definition 5.2 ω is called a *balanced metric* for $L^{\otimes\gamma}$ if $|v_1|_h^2 + |v_2|_h^2 + \cdots + |v_N|_h^2$ is a constant function on X for an orthonormal basis $\{v_1, v_2, \cdots, v_N\}$ of (V_γ, ρ_γ).

Theorem 5.1 ([93]; See Also [40]) $X \subset \mathbb{P}(V_\gamma^*)$ *is Chow polystable if and only if there exists a balanced metric ω for $L^{\otimes\gamma}$.*

Proof We first use the following Hilbert–Mumford stability criterion (see for instance [48]): *A G-orbit in W is closed if and only if for some point w in the orbit, $\sigma(\mathbb{C}^*) \cdot w$ is closed in W for every special one-parameter group $\sigma : \mathbb{C}^* \to G$.*

Then for the "if" part of the proof, it suffices to show that $\sigma(\mathbb{C}^*) \cdot \hat{X}$ is closed for every special one-parameter group $\sigma : \mathbb{C}^* \to G$ by assuming that a balanced metric ω for $L^{\otimes\gamma}$ exists. Then by identifying V^* with $\mathbb{C}^N = \{(z_1, \cdots, z_N)\}$ by a suitable choice of an orthonormal basis $\{v_1, \ldots, v_N\}$ of V, we can diagonalize σ in the form

$$\sigma(t) \cdot z = (t^{b_1} z_1, \cdots, t^{b_n} z_N),$$

where $z = (z_1, \cdots, z_N) \in V^*$, and $-b_i \in \mathbb{Z}$ are the weights of the $\mathbb{C}^*$-action on V via σ. Since $G = \mathrm{SL}(V)$, it follows that $\sum_{i=1}^N b_i = 0$. Note also that

$$|v_1|_h^2 + \cdots + |v_N|_h^2 = C \tag{5.1}$$

for some positive constant C. For the balanced metric ω, we have the Hermitian structure ρ_γ for $V = V_\gamma$. Put $f(s) := \log \|\sigma(e^s) \cdot \hat{X}\|_{\mathrm{CH}(\rho_\gamma)}$ for $s \in \mathbb{R}$. Then by (1.8),

$$\dot{f}(0) = (n+1) \int_X \frac{b_1 |v_1|_h^2 + \cdots + b_N |v_N|_h^2}{|v_1|_h^2 + \cdots + |v_N|_h^2} \, \omega_{\mathrm{FS}}^n, \tag{5.2}$$

where $\omega_{\mathrm{FS}} := dd^c \log(\sum_{i=1}^n |v_i|^2)$. On the other hand, let $dd^c \log$ operate on both sides of (5.1). It then follows that

$$\omega_{\mathrm{FS}} = -\gamma \, dd^c \log h = \gamma\omega. \tag{5.3}$$

Now by (5.1)–(5.3), we obtain

$$
\dot{f}(0) \;=\; \frac{(n+1)\gamma^n}{C} \sum_{i=1}^{N} b_i \int_X |v_i|_h^2 \omega^n \;=\; \frac{(n+1)\gamma^n}{C} \sum_{i=1}^{N} b_i \;=\; 0. \qquad (5.4)
$$

Moreover, by the second variation formula for the Chow norm, $\ddot{f}(0) \geq 0$. On the other hand, for all $t \in \mathbb{C}^*$, we have

$$
\|\sigma(t) \cdot \hat{X}\|_{\mathrm{CH}(\rho_\gamma)} = \|\sigma(|t|) \cdot \hat{X}\|_{\mathrm{CH}(\rho_\gamma)}, \qquad (5.5)
$$

since for the subgroup S^1 of $\mathbb{C}^*$, its image $\sigma(S^1)$ acts isometrically on (V_γ, ρ_γ) and hence on $(W, \mathrm{CH}(\rho_\gamma))$. Then the following cases are possible:

Case 1: $\ddot{f}(0) > 0$. Note that the function $f(s)$ is convex (see Sect. 1.4). We now see from (5.4) and the inequality $\ddot{f}(0) > 0$ that

$$
\lim_{s \to -\infty} f(s) \;=\; +\infty \;=\; \lim_{s \to +\infty} f(s).
$$

Hence by (5.5), in this case, $\sigma(\mathbb{C}^*) \cdot \hat{X}$ is closed in W.

Case 2: $\ddot{f}(0) = 0$. Then by Problem 1.1, the one-parameter group $\sigma : \mathbb{C}^* \to G$ preserves the subvariety X in $\mathbb{P}(V^*)$. Hence there exists an integer α such that

$$
\sigma(t) \cdot \hat{X} \;=\; t^\alpha \hat{X}, \qquad t \in \mathbb{C}^*.
$$

Then by setting $t = e^s$ with $s \in \mathbb{R}$, we have

$$
0 \;=\; \dot{f}(0) \;=\; \frac{d(\log \|\sigma(t) \cdot \hat{X}\|_{\mathrm{CH}(\rho_\gamma)})}{ds}\bigg|_{s=0} \;=\; \frac{d(\log \|e^{s\alpha} \cdot \hat{X}\|_{\mathrm{CH}(\rho_\gamma)})}{ds}\bigg|_{s=0}
$$

$$
\;=\; \frac{d(\alpha s + \log \|\hat{X}\|_{\mathrm{CH}(\rho_\gamma)})}{ds}\bigg|_{s=0} \;=\; \alpha.
$$

Hence $\sigma(\mathbb{C}^*) \cdot \hat{X}$ is a single point $\hat{X}$, and in particular is closed.

For the "only if" part of the proof, assume that $X \subset \mathbb{P}(V_\gamma^*)$ is Chow polystable, i.e., $G \cdot \hat{X}$ is closed in W. For a Kähler metric $\omega = \mathrm{Ric}(h)$, the associated Chow norm

$$
G \cdot \hat{X} \ni w \mapsto \|w\|_{\mathrm{CH}(\rho_\gamma)} \in \mathbb{R}_{\geq 0}
$$

attains its minimum at some point $g_0 \cdot \hat{X}$ in the orbit (where $g_0 \in G$). Put $K := \mathrm{SU}(V_\gamma, \rho_\gamma)$. By choosing an orthonormal basis $\{v_1, \cdots, v_N\}$ for V_γ, we can identify $G = \mathrm{SL}(V_\gamma)$ with $\mathrm{SL}(N, \mathbb{C})$. Then for some $k, k_1 \in K$, we can write

$$g_0 = k \, \Delta_0 \, k_1,$$

where Δ_0 is a positive real diagonal matrix of order N whose α-th diagonal element is a_α^{-1}. Since the Chow norm $\| \ \|_{\mathrm{CH}(\rho_\gamma)}$ is K-invariant, k in the above expression of g_0 can be chosen to be arbitrary in K. Let $1 \leq \alpha_1 < \alpha_2 \leq N$. Then by the minimality of the Chow norm at $g_0 \cdot \hat{X}$, we obtain

$$(d/ds)_{|s=0} \| \exp(sv) \cdot k \, \Delta_0 \, k_1 \cdot \hat{X} \|_{\mathrm{CH}(\rho_\gamma)} = 0, \quad v \in \mathfrak{g} := \mathfrak{sl}(N, \mathbb{C}). \tag{5.6}$$

First, by rotating the Kodaira embedding by k_1^{-1}, we may replace $k_1 \cdot \hat{X}$ by $\hat{X}$, so that we may assume $k_1 = 1$ from the beginning. Next, let v be the diagonal matrix in $\mathfrak{g}$ whose α-th diagonal element is $\delta_{\alpha\alpha_1} - \delta_{\alpha\alpha_2}$. Here $\delta_{\alpha\beta}$ is Kronecker's delta, so that $\delta_{\alpha\beta} = 1$ or 0, if $\alpha = \beta$ or $\alpha \neq \beta$, respectively. Then by (1.8) and (5.6) applied to $k = 1$, we obtain

$$\int_X \frac{|z'_{\alpha_1}|^2}{\sum_{\alpha=1}^N |z'_\alpha|^2} \, {\omega'_{\mathrm{FS}}}^n = \int_X \frac{|z'_{\alpha_2}|^2}{\sum_{\alpha=1}^N |z'_\alpha|^2} \, {\omega'_{\mathrm{FS}}}^n, \tag{5.7}$$

where $z'_\alpha := a_\alpha z_\alpha$ and $\omega'_{\mathrm{FS}} := dd^c \log(\sum_{\alpha=1}^N |z'_\alpha|^2)$. Let $\{v'_1, \cdots, v'_N\}$ be the basis for V_γ associated to the system of coordinates $z' = {}^t(z'_1, \cdots, z'_N)$, i.e., $v'_\alpha := a_\alpha v_\alpha$, $\alpha = 1, 2, \cdots, N$. Then (5.7) is written as

$$\int_X \frac{|v'_{\alpha_1}|^2}{\sum_{\alpha=1}^N |v'_\alpha|^2} \, {\omega'_{\mathrm{FS}}}^n = \int_X \frac{|v'_{\alpha_2}|^2}{\sum_{\alpha=1}^N |v'_\alpha|^2} \, {\omega'_{\mathrm{FS}}}^n = C, \tag{5.8}$$

where C is a positive constant independent of the choice of α_1 and α_2. Next, let $k_2 \in K$ be such that $k_2 z = {}^t((k_2 z)_1, \cdots, (k_2 z)_N)$ for all $z \in \mathbb{C}^N$, where

$$(k_2 z)_\alpha = \begin{cases} (1/\sqrt{2})(z_{\alpha_1} - z_{\alpha_2}), & \alpha = \alpha_1, \\ (1/\sqrt{2})(z_{\alpha_1} + z_{\alpha_2}), & \alpha = \alpha_2, \\ z_\alpha, & \alpha_1 \neq \alpha \neq \alpha_2. \end{cases}$$

Moreover, let $k_3 \in K$ be such that $k_3 z = ((k_3 z)_1, \cdots, (k_3 z)_N)$ for all $z \in \mathbb{C}^N$, where

$$(k_3 z)_\alpha = \begin{cases} (1/\sqrt{2})(z_{\alpha_1} + \sqrt{-1}\, z_{\alpha_2}), & \alpha = \alpha_1, \\ (1/\sqrt{2})(\sqrt{-1}\, z_{\alpha_1} + z_{\alpha_2}), & \alpha = \alpha_2, \\ z_\alpha, & \alpha_1 \neq \alpha \neq \alpha_2. \end{cases}$$

Then by a straightforward computation, we obtain

$$-\{\,|(k_2 z')_{\alpha_1}|^2 - |(k_2 z')_{\alpha_2}|^2\,\} + \sqrt{-1}\{\,|(k_3 z')_{\alpha_1}|^2 - |(k_3 z')_{\alpha_2}|^2\,\} = 2z'_{\alpha_1}\bar{z}'_{\alpha_2}. \tag{5.9}$$

By $k_2 \in K$, both $\sum_{\alpha=1}^{N} |(k_2 z')_\alpha|^2 = \sum_{\alpha=1}^{N} |z'_\alpha|^2$ and $\omega'_{FS} = dd^c \log \sum_{\alpha=1}^{N} |(k_2 z')_\alpha|^2$ hold, and the same thing is true also for k_3. As we obtain the equality (5.7) from (5.6) applied to $k = 1$, we obtain the following from (5.6) applied to $k = k_2$:

$$\int_X \frac{|(k_2 z')_{\alpha_1}|^2}{\sum_{\alpha=1}^{N} |z'_\alpha|^2} \omega'_{FS}{}^n = \int_X \frac{|(k_2 z')_{\alpha_2}|^2}{\sum_{\alpha=1}^{N} |z'_\alpha|^2} \omega'_{FS}{}^n. \tag{5.10}$$

Similarly, by applying (5.6) to $k = k_3$, we obtain the following:

$$\int_X \frac{|(k_3 z')_{\alpha_1}|^2}{\sum_{\alpha=1}^{N} |z'_\alpha|^2} \omega'_{FS}{}^n = \int_X \frac{|(k_3 z')_{\alpha_2}|^2}{\sum_{\alpha=1}^{N} |z'_\alpha|^2} \omega'_{FS}{}^n. \tag{5.11}$$

Then by (5.9)–(5.11),

$$\int_X \frac{v'_{\alpha_1}\bar{v}'_{\alpha_2}}{\sum_{\alpha=1}^{N} |v'_\alpha|^2} \omega'_{FS}{}^n = \int_X \frac{z'_{\alpha_1}\bar{z}'_{\alpha_2}}{\sum_{\alpha=1}^{N} |z'_\alpha|^2} \omega'_{FS}{}^n = 0. \tag{5.12}$$

Let u be a local section for $L^{\otimes\gamma}$. Then

$$|u|^2_{h'_{FS}} := \frac{|u|^2}{\sum_{\alpha=1}^{N} |v'_\alpha|^2}$$

defines a Hermitian metric for $L^{\otimes\gamma}$, so that $h' := (h'_{FS})^{1/\gamma}$ is a Hermitian metric for L. The corresponding Ricci form

$$\omega' := \text{Ric}(h') = -dd^c \log h' = \frac{1}{\gamma}dd^c \log\left(\sum_{\alpha=1}^{n} |v'_\alpha|^2\right) = \frac{\omega'_{FS}}{\gamma} \tag{5.13}$$

is a Kähler form on X. As in (2.2), let ρ' be the Hermitian structure for V_γ induced by h' and ω' such that

$$\langle v_1, v_2 \rangle_{\rho'} := \int_X (v_1, v_2)_{h'}\, \omega'^n, \qquad v_1, v_2 \in V_\gamma.$$

Put $u_\alpha := (\gamma^n/C)^{1/2} v'_\alpha$. In view of (5.13), we see from (5.8) and (5.12) that $\{u_\alpha \, ; \, \alpha = 1, 2, \ldots, N\}$ is an orthonormal basis for (V_γ, ρ'). Then

$$|u_1|^2_{h'} + \cdots + |u_N|^2_{h'} = |u_1|^2_{h'_{FS}} + \cdots + |u_N|^2_{h'_{FS}} = (\gamma^n/C)(|v'_1|^2_{h'_{FS}} + \cdots + |v'_N|^2_{h'_{FS}})$$

$$= (\gamma^n/C) \left(\frac{|v'_1|^2}{\sum_{\alpha=1}^N |v'_\alpha|^2} + \cdots + \frac{|v'_N|^2}{\sum_{\alpha=1}^N |v'_\alpha|^2} \right) = \gamma^n/C,$$

and therefore ω' is a balanced metric, as required. $\square$

For a special one-parameter group $\sigma : \mathbb{C}^* \to G$, we consider the real-valued function $f_\sigma(s) := \log \|\sigma(e^s) \cdot \hat{X}\|_{\mathrm{CH}(\rho_\gamma)}$, $s \in \mathbb{R}$, where $G = \mathrm{SL}(V)$ for $V := V_\gamma$. Then by the same argument as in obtaining (2.12) in Theorem 2.2,

$$\lambda_\sigma = \lim_{s \to -\infty} \dot{f}_\sigma(s), \tag{5.14}$$

where the left-hand side λ_σ is the *Chow weight* for $X \subset \mathbb{P}(V^*_\gamma)$ for the $\mathbb{C}^*$-action via σ, i.e., the weight of the $\mathbb{C}^*$-action via σ on the complex line in W associated to $\lim_{|t| \to 0} [\sigma(t)\hat{X}]$. For the special one-parameter group

$$\sigma^{-1} : \mathbb{C}^* \to G$$

defined by $\sigma^{-1}(t) := \sigma(t)^{-1} = \sigma(t^{-1})$, $t \in \mathbb{C}^*$, we consider the real-valued function $f_{\sigma^{-1}}(s) := \log \|\sigma^{-1}(e^s) \cdot \hat{X}\|_{\mathrm{CH}(\rho_\gamma)}$ on $\mathbb{R}$. Then

$$\lambda_{\sigma^{-1}} = \lim_{s \to -\infty} \dot{f}_{\sigma^{-1}}(s),$$

where $\lambda_{\sigma^{-1}}$ is the Chow weight for $X \subset \mathbb{P}(V^*_\gamma)$ for the $\mathbb{C}^*$-action via σ^{-1}. Since $f_{\sigma^{-1}}(s) = f_\sigma(-s)$ for all $s \in \mathbb{R}$, we obtain

$$\lambda_{\sigma^{-1}} = - \lim_{s \to +\infty} \dot{f}_\sigma(s). \tag{5.15}$$

Theorem 5.2 *For a special one-parameter group $\sigma : \mathbb{C}^* \to G = \mathrm{SL}(V_\gamma)$, the orbit $\sigma(\mathbb{C}^*) \cdot \hat{X}$ is closed in W if and only if both $\lambda_\sigma < 0$ and $\lambda_{\sigma^{-1}} < 0$ hold.*

Proof By the second variation formula for the Chow norm (see Sect. 1.4), the real-valued function $f_\sigma(s)$ on $\mathbb{R}$ is convex. In view of (5.14) and (5.15), it is easy to check the following equivalence:

$$\sigma(\mathbb{C}^*) \cdot \hat{X} \text{ is closed} \iff \lim_{s \to -\infty} f_\sigma(s) = +\infty = \lim_{s \to +\infty} f_\sigma(s)$$

$$\iff \lambda_\sigma < 0 \text{ and } \lambda_{\sigma^{-1}} < 0.$$
$\square$

Corollary 5.1 $X \subset \mathbb{P}(V_\gamma^*)$ *is Chow polystable if* $\lambda_\sigma < 0$ *for every nontrivial special one-parameter group* $\sigma : \mathbb{C}^* \to G$ *for* $G = \mathrm{SL}(V_\gamma)$.

Proof By the Hilbert–Mumford stability criterion as in the proof of Theorem 5.1, this corollary is straightforward from Theorem 5.2 above. $\quad\square$

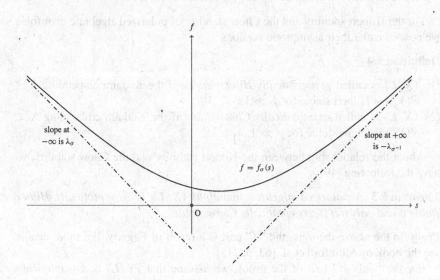

5.2 The Hilbert Stability

For an n-dimensional polarized algebraic manifold (X, L), we set $V_\gamma := H^0(X, L^{\otimes\gamma})$ and $G_\gamma := \mathrm{SL}(V_\gamma)$ for a positive integer γ. We further put $V_{k\gamma} := H^0(X, L^{\otimes k\gamma})$ for positive integers k. By choosing $\gamma \gg 1$, we may assume that the natural maps

$$\Psi_{k,\gamma} : S^k(V_\gamma) \to V_{k\gamma}, \qquad k = 1, 2, \ldots,$$

are surjective (cf. [61]), where $S^k(V_\gamma)$ denotes the k-th symmetric tensor product of V_γ. We now put $I_{k,\gamma} := \mathrm{Ker}\, \Psi_{k,\gamma}$ and $n_{k,\gamma} := \dim I_{k,\gamma}$. Then

$$\ell_{k,\gamma} := \det I_{k,\gamma} = \wedge^{n_{k,\gamma}} I_{k,\gamma}$$

is a complex line sitting in $\wedge^{n_{k,\gamma}} S^k(V_\gamma)$. Let $0 \neq p_{k,\gamma} \in \ell_{k,\gamma}$. We further consider the Kodaira embedding $X \subset \mathbb{P}(V_\gamma^*)$ associated to the complete linear system $|L^{\otimes\gamma}|$.

Definition 5.3

(1) $X \subset \mathbb{P}(V_\gamma^*)$ is called *Hilbert polystable* if the orbit $G_\gamma \cdot p_{k,\gamma}$ is closed in the vector space $\wedge^{n_{k,\gamma}} S^k(V_\gamma)$ for $k \gg 1$.

(2) Let $X \subset \mathbb{P}(V_\gamma^*)$ be Hilbert polystable. Then $X \subset \mathbb{P}(V_\gamma^*)$ is called *Hilbert stable* if the isotropy subgroup of G_γ at $p_{k,\gamma}$ is finite for $k \gg 1$.

For the Hilbert stability and the Chow stability of polarized algebraic manifolds, we now consider their asymptotic versions.

Definition 5.4

(1) (X, L) is called *asymptotically Hilbert stable* if the Kodaira embedding $X \subset \mathbb{P}(V_\gamma^*)$ is Hilbert stable for $\gamma \gg 1$.

(2) (X, L) is called *asymptotically Chow stable* if the Kodaira embedding $X \subset \mathbb{P}(V_\gamma^*)$ is Chow stable for $\gamma \gg 1$.

About the relationship between the Hilbert stability and the Chow stability, we have the following [49]:

Theorem 5.3 *A polarized algebraic manifold (X, L) is asymptotically Hilbert stable if and only if it is asymptotically Chow stable.*

Proof In the above theorem, the "if" part is a result of Fogarty. For more details, see the book by Mumford et al. [63, p. 215].

For the "only if" part of the proof, we assume that (X, L) is asymptotically Hilbert stable. For the Kodaira embedding

$$\Phi_\gamma : X \hookrightarrow \mathbb{P}(V_\gamma^*)$$

associated to the complete linear system $|L^{\otimes \gamma}|$ on X, we consider its image $X_\gamma := \Phi_\gamma(X)$ and the degree $d_\gamma := \deg_{\mathbb{P}(V_\gamma^*)} X_\gamma$. Let $0 \neq \hat{X}_\gamma \in W_\gamma := (\mathrm{Sym}^{d_\gamma} V_\gamma^*)^{\otimes n+1}$ be the Chow form for the irreducible reduced algebraic cycle X_γ. Since it is a routine task to show the finiteness of the isotropy subgroup of $G_\gamma := \mathrm{SL}(V_\gamma)$ at $\hat{X}_\gamma$ for $\gamma \gg 1$ from the finiteness of the isotropy subgroup of G_γ at $p_{k,\gamma}$ for $k \gg 1$, the proof of the "only if" part is reduced to showing that $G_\gamma \cdot \hat{X}_\gamma$ is closed in W_γ for $\gamma \gg 1$. Then by the Hilbert–Mumford stability criterion (as in the proof of Theorem 5.1), it suffices to show that the orbit $\sigma(\mathbb{C}^*) \cdot \hat{X}_\gamma$ is closed in W_γ for every special one-parameter group $\sigma : \mathbb{C}^* \to G_\gamma$, provided that $\gamma \gg 1$.

In order to show this, given a nontrivial special one-parameter group $\sigma : \mathbb{C}^* \to G_\gamma$, we shall define a sequence of special one-parameter groups

$$\sigma_\alpha : \mathbb{C}^* \to \mathrm{GL}(V_{\gamma_\alpha}), \qquad \alpha = 1, 2, \ldots,$$

such that each γ_α is a multiple of $\gamma_{\alpha-1}$, where $\gamma_0 := \gamma \gg 1$. Let $(\mathscr{X}_\sigma, \mathscr{L}_\sigma)$ be the test configuration, of exponent γ, for (X, L) associated to the one-parameter group σ above, where $\pi : \mathscr{X}_\sigma \to \mathbb{A}^1$ is the natural projection. Put $\gamma_\alpha := k_\alpha \gamma$ for a positive integer k_α written in the form

$$k_\alpha = \prod_{i=1}^{\alpha} K_i \gg 1, \qquad \alpha = 1, 2, \ldots,$$

where all K_i, $i = 1, 2, \ldots, \alpha$, are integers satisfying $K_i \gg 1$. Then $\gamma_\alpha = K_\alpha \gamma_{\alpha-1}$, and hence γ_α is a multiple of $\gamma_{\alpha-1}$. For the direct image sheaves

$$\mathscr{E}^{(\alpha)} := \pi_* \mathscr{L}_\sigma^{\otimes k_\alpha}, \qquad \alpha = 1, 2, \cdots,$$

the affirmative solution of the equivariant Serre conjecture for abelian groups gives us a $\mathbb{C}^*$-equivariant isomorphism $\mathscr{E}^{(\alpha)} \cong \mathscr{E}_0^{(\alpha)} \times \mathbb{A}^1$, so that we have an identification

$$\mathscr{E}_0^{(\alpha)} \cong \mathscr{E}_1^{(\alpha)} = V_{\gamma_\alpha}, \tag{5.16}$$

by an argument as in Sect. 1.1. Since $E_\alpha := H^0((\mathscr{X}_\sigma)_0, (\mathscr{L}_\sigma^{\otimes k_\alpha})_0)$ and $\mathscr{E}_0^{(\alpha)}$ are $\mathbb{C}^*$-equivariantly identified, in view of (5.16), the natural $\mathbb{C}^*$-action on $\mathscr{E}_0^{(\alpha)}$ induces a special one-parameter group

$$\sigma_\alpha : \mathbb{C}^* \to GL(E_\alpha) \ (= GL(V_{\gamma_\alpha})),$$

so that we can view $(\mathscr{X}_\sigma, \mathscr{L}_\sigma^{\otimes k_\alpha})$ as the test configuration $(\mathscr{X}_{\sigma_\alpha}, \mathscr{L}_{\sigma_\alpha})$ associated to σ_α. As in the proof of Theorem 2.2, by Mumford [62, Proposition 2.11], the weight w_{γ_α} of the $\mathbb{C}^*$-action on $\det E_\alpha$ via σ_α is

$$w_{\gamma_\alpha} = -\frac{\lambda_\sigma}{(n+1)!} k_\alpha^{n+1} + O(k_\alpha^n),$$

where λ_σ is the Chow weight of $X \subset \mathbb{P}(V_\gamma^*)$ for the $\mathbb{C}^*$-action via σ. Since k_α tends to $+\infty$ as $\alpha \to \infty$, and since we can write $N_{\gamma_\alpha} := \dim V_{\gamma_\alpha}$ as a polynomial of γ_α by the Riemann–Roch theorem, it follows from the equality $\gamma_\alpha = k_\alpha \gamma$ that

$$\lim_{\alpha \to \infty} \frac{w_{\gamma_\alpha}}{\gamma_\alpha N_{\gamma_\alpha}} = \lim_{\alpha \to \infty} \frac{-\frac{\lambda_\sigma}{(n+1)!} k_\alpha^{n+1} + O(k_\alpha^n)}{k_\alpha \gamma \left\{ \frac{c_1(L)^n[X]}{n!} (k_\alpha \gamma)^n + O((k_\alpha \gamma)^{n-1}) \right\}}$$

$$= -\frac{\lambda_\sigma}{(n+1)\gamma^{n+1} c_1(L)^n[X]}.$$

Hence by Corollary 5.1, it suffices to show the positivity of the limit on the left-hand side for every special one-parameter group $\sigma : \mathbb{C}^* \to G$.

In order to show such positivity, by setting $q_\alpha := w_{\gamma_\alpha}/(\gamma_\alpha N_{\gamma_\alpha})$, we claim that the sequence $\{q_\alpha\}$ is monotone-increasing, i.e.,

$$q_{\alpha-1} < q_\alpha, \qquad \alpha = 1, 2, \cdots. \qquad (5.17)$$

From now on, K_α will be written simply as κ. Then we can write γ_α as $\kappa\gamma_{\alpha-1}$. Let $\psi_\alpha : S^\kappa(E_{\alpha-1}) \to E_\alpha$ be the natural map on the special fiber (over the origin). Then by $\gamma_{\alpha-1} \gg 1$, in view of [61] and [62], ψ_α is surjective. Moreover, by (5.16), $E_{\alpha-1}$ and E_α are viewed as $V_{\gamma_{\alpha-1}}$ and V_{γ_α}, respectively. Put $\iota_\alpha := \mathrm{Ker}\,\psi_\alpha$. Then the weight H_α, which we call the Hilbert weight, of the $\mathbb{C}^*$-action on $\det \iota_\alpha$ is (see [20])

$$H_\alpha = \{\mathbb{C}^*\text{-weight on } \det S^\kappa(E_{\alpha-1})\} - (\mathbb{C}^*\text{-weight on } \det E_\alpha)$$
$$= \frac{\kappa\, w_{\gamma_{\alpha-1}}}{N_{\gamma_{\alpha-1}}} N_{\gamma_\alpha} - w_{\gamma_\alpha}.$$

In view of $\gamma_{\alpha-1} \gg 1$, the closeness of the $\mathbb{C}^*$-orbit $\sigma_\alpha(\mathbb{C}^*) \cdot p_{\kappa,\gamma_{\alpha-1}}$ in W_{γ_α} implies that the Hilbert weight H_α is negative, i.e.,

$$0 > \frac{\kappa\, w_{\gamma_{\alpha-1}}}{N_{\gamma_{\alpha-1}}} N_{\gamma_\alpha} - w_{\gamma_\alpha}.$$

Since $\gamma_\alpha = \kappa\gamma_{\alpha-1}$, dividing this inequality by $\gamma_\alpha N_{\gamma_\alpha}$, we obtain (5.17). Note that $q_0 = w_\gamma/(\gamma N_\gamma) = 0$, since we start from $\sigma : \mathbb{C}^* \to G = \mathrm{SL}(V_\gamma)$. Hence

$$0 = q_0 < q_1 < \cdots < q_\alpha < \cdots,$$

and we obtain $\lim_{\alpha\to\infty} q_\alpha > 0$, as required.   □

5.3 K-stability

The concept of K-stability was introduced by Tian [79] in his study of Kähler–Einstein metrics on Fano manifolds. Later, by Donaldson [20], this concept was reformulated in a general setting by an algebraic geometric language. We consider here an n-dimensional polarized algebraic manifold (X, L). A test configuration $(\mathscr{X}, \mathscr{L})$ for (X, L) is called *normal* if $\mathscr{X}$ is a normal complex variety.

Definition 5.5

(1) (X, L) is called *K-semistable* if $\mathrm{DF}_1(\mathscr{X}, \mathscr{L}) \leq 0$ for every normal test configuration $(\mathscr{X}, \mathscr{L})$ for (X, L).

(2) (X, L) is called *K-polystable* if (X, L) is K-semistable and furthermore, every normal test configuration $(\mathscr{X}, \mathscr{L})$ satisfying $\mathrm{DF}_1(\mathscr{X}, \mathscr{L}) = 0$ is a product configuration.

(3) (X, L) is called *K-stable* if (X, L) is K-semistable and furthermore, every normal test configuration $(\mathscr{X}, \mathscr{L})$ satisfying $\mathrm{DF}_1(\mathscr{X}, \mathscr{L}) = 0$ is trivial.

Definition 5.6

(1) (X, L) is called *uniformly K-stable* if every normal test configuration $\mu = (\mathscr{X}, \mathscr{L})$ for (X, L) satisfies $\mathrm{DF}_1(\mu) \leq -C\|\mu\|_{asymp}$ for some positive constant C independent of the choice of μ.
(2) (X, L) is called *strongly K-stable* if $F_1(\{\mu_j\}) < 0$ for all $\{\mu_j\} \in \mathscr{M}$, where $F_1(\{\mu_j\})$ and $\mathscr{M}$ are as in Sect. 4.5.

5.4 Relative Stability

As a reference for this section, see [76] (cf. also [55]). For a polarized algebraic manifold (X, L), if $\dim H^0(X, \mathscr{O}(TX)) \neq 0$, then for the Kähler class $c_1(L)$ on X to admit a special metric such as a CSC Kähler metric or an extremal Kähler metric, the stability concept should be modified suitably by choosing a smaller reductive algebraic group. In this section, we consider a possibly trivial algebraic torus T sitting in $\mathrm{Aut}^0(X)$. Then for every positive integer γ, we consider the reductive algebraic subgroup $T_\gamma^\perp$, as in Sect. 4.3, of the full special linear group $G_\gamma := \mathrm{SL}(V_\gamma)$. As in Sect. 5.1, we consider the abstract Kodaira embedding

$$\Phi_\gamma : X \hookrightarrow \mathbb{P}(V_\gamma^*)$$

associated to $|L^{\otimes\gamma}|$. Let $X_\gamma := \Phi_\gamma(X)$ be its image. As in the proof of Theorem 5.3, let $0 \neq \hat{X}_\gamma \in W_\gamma := (\mathrm{Sym}^{d_\gamma} V_\gamma^*)^{\otimes n+1}$ be the Chow form for the cycle X_γ on $\mathbb{P}^*(V_\gamma)$.

Definition 5.7

(1) $X \subset \mathbb{P}(V_\gamma^*)$ is called *Chow polystable relative to T* if the orbit $T_\gamma^\perp \cdot \hat{X}_\gamma$ is closed in W_γ.
(2) $X \subset \mathbb{P}(V_\gamma^*)$ is called *Chow stable relative to T* if $X \subset \mathbb{P}(V_\gamma^*)$ is Chow polystable relative to T and in addition, the isotropy subgroup of $T_\gamma^\perp$ at $\hat{X}_\gamma$ is finite.
(3) (X, L) is called *asymptotically Chow polystable relative to T* if $X \subset \mathbb{P}(V_\gamma^*)$ is Chow polystable relative to T for $\gamma \gg 1$.
(4) (X, L) is called *asymptotically Chow stable relative to T* if $X \subset \mathbb{P}(V_\gamma^*)$ is Chow stable relative to T for $\gamma \gg 1$.

For each $\sigma \in \mathrm{1PS}(T_\gamma^\perp)$, let $u \in \mathfrak{t}_\gamma^\perp$ be its fundamental generator. As in Sect. 2.2, we have the test configuration $\mu_\sigma = (\mathscr{X}_\sigma, \mathscr{L}_\sigma)$ associated to σ. For positive integers m, we put $\gamma_m := m\gamma$. As in Sect. 4.2, u induces

$$u_m \in \mathfrak{sl}(H^0((\mathscr{X}_\sigma)_0, (\mathscr{L}_\sigma^{\otimes m})_0)).$$

Let v_m be as in (4.3), where $v \in \mathfrak{t}_\mathbb{R}$ is chosen arbitrarily. Then by Székelyhidi [76], the pairing $\langle \ , \ \rangle_{\gamma_m}$ in (4.2) converges to a limit pairing $\langle \ , \ \rangle_\infty$ on $\mathfrak{z}_\gamma \times \mathfrak{t}$ such that

$$\langle u_m, v_m \rangle_{\gamma_m} \to \langle u, v \rangle_\infty, \quad \text{as } m \to \infty.$$

Let w_i, $i = 1, \ldots, r$, be elements in $\mathfrak{t}$ with real eigenvalues such that $\{w_1, \cdots, w_r\}$ is an orthonormal basis for $\mathfrak{t}$ $(= \mathfrak{t}_\gamma)$ with respect to $\langle \ , \ \rangle_\infty$. Then we put

$$u' := u - \sum_{i=1}^{r} \langle u, w_i \rangle_\infty \, w_i,$$

where u' is orthogonal to $\mathfrak{t}$ in terms of the bilinear pairing $\langle \ , \ \rangle_\infty$. We now define the relative Donaldson-Futaki invariant DF_1^T (cf. [76]) for $(\mathscr{X}_\sigma, \mathscr{L}_\sigma)$ by

$$\mathrm{DF}_1^T (\mathscr{X}_\sigma, \mathscr{L}_\sigma) := \mathrm{DF}_1(u').$$

Definition 5.8

(1) (X, L) is called *K-semistable relative to T* if for every positive integer γ, $\mathrm{DF}_1^T(\mathscr{X}_\sigma, \mathscr{L}_\sigma) \leq 0$ for all $\sigma \in 1\mathrm{PS}(T_\gamma^\perp)$ such that $\mathscr{X}_\sigma$ is normal.

(2) (X, L) is called *K-stable relative to T* if for every positive integer γ, the inequality $\mathrm{DF}_1^T(\mathscr{X}_\sigma, \mathscr{L}_\sigma) < 0$ holds for all $\sigma \in 1\mathrm{PS}(T_\gamma^\perp)$ as long as $\mathscr{X}_\sigma$ is normal.

(3) (X, L) is called *K-polystable relative to T* if for every positive integer γ and for every $\sigma \in 1\mathrm{PS}(T_\gamma^\perp)$, the inequality $\mathrm{DF}_1^T(\mathscr{X}_\sigma, \mathscr{L}_\sigma) \leq 0$ holds and, as long as $\mathscr{X}_\sigma$ is normal, the equality holds only when $(\mathscr{X}_\sigma, \mathscr{L}_\sigma)$ is a product configuration.

Definition 5.9

(1) (X, L) is called *uniformly K-stable relative to T* if for all positive integers γ and all $\sigma \in 1\mathrm{PS}(T_\gamma^\perp)$ such that $\mathscr{X}_\sigma$ is a normal complex variety, there exists a positive constant C independent of the choice of γ and σ such that the inequality $\mathrm{DF}_1^T(\mu_\sigma) \leq -C\|\mu_\sigma\|_{asymp}^T$ holds for $\mu_\sigma = (\mathscr{X}_\sigma, \mathscr{L}_\sigma)$.

(2) (X, L) is *strongly K-semistable relative to T* if $F_1(\{\mu_j\}) \leq 0$ for all $\{\mu_j\} \in \mathscr{M}_T$.

(3) (X, L) is *strongly K-stable relative to T* if $F_1(\{\mu_j\}) < 0$ for all $\{\mu_j\} \in \mathscr{M}_T$.

If the algebraic torus T is trivial, then all relative stabilities reduce to the ordinary stabilities without the term "relative to T". We now compare strong K-stability with uniform K-stability by assuming that T is trivial for simplicity, though a similar argument goes through also for a relative version.

Theorem 5.4 *If (X, L) is strongly K-stable, then (X, L) is uniformly K-stable.*

Proof By choosing a Hermitian metric h for L such that $\omega := \mathrm{Ric}(h)$ is Kähler, we can talk about special one-parameter groups. Let (X, L) be strongly K-stable. Since $\mathscr{X}_\sigma$ is assumed to be normal for σ above, by the nontriviality of σ, we have the inequality $\|\mu_\sigma\|_{asymp} > 0$. In general, for every nontrivial normal test configuration $\mu = (\mathscr{X}, \mathscr{L})$ for (X, L), we put

$$\kappa(\mu) := \frac{\mathrm{DF}_1(\mu)}{\|\mu\|_{asymp}}.$$

Since the uniform stability means that $\kappa(\sigma)$ is bounded from above by $-C$, we assume for contradiction that there exist positive integers γ_j, $j = 1, 2, \cdots$, and nontrivial special one-parameter groups $\sigma_j : \mathbb{C}^* \to \mathrm{SL}(V_{\gamma_j})$ satisfying the following:

(1) $\kappa(\mu_1) \leq \cdots \leq \kappa(\mu_j) \leq \kappa(\mu_{j+1}) \leq \cdots$,
(2) $\lim_{j\to\infty} \kappa(\mu_j) \geq 0$,

where μ_j is the test configuration $(\mathscr{X}_{\sigma_j}, \mathscr{L}_{\sigma_j})$ associated to σ_j, and the limit in (2) can possibly be $+\infty$. Then for the test configuration

$$\mu_{j,m} := (\mathscr{X}_j, \mathscr{L}_j^{\otimes m}), \qquad m \gg 1,$$

the $\mathbb{C}^*$-action on the central fiber via σ_j induces $\sigma_{j,m} : \mathbb{C}^* \to \mathrm{GL}(\mathscr{E}_{j,m})$, where the space $\mathscr{E}_{j,m} := H^0((\mathscr{X}_j)_0, (\mathscr{L}_j)_0^{\otimes m})$ is viewed as $V_{m\gamma_j}$. As in Sect. 2.5, we have

$$\sigma_{j,m}^{\mathrm{SL}} : \mathbb{R}_+ \to \mathrm{SL}(\mathscr{E}_m).$$

Let $u_{j,m} \in \mathfrak{sl}(V_{m\gamma_j})$ be the fundamental generator for $\sigma_{j,m}^{\mathrm{SL}}$. Then by setting $t = \exp(s/|u_{j,m}|_\infty)$, we define a real-valued function $f_{j,m}(s)$ by

$$f_{j,m}(s) := \frac{|u_{j,m}|_\infty}{|u_{j,m}|_1} (m\gamma_j)^{-n} \|\sigma_{j,m}^{\mathrm{SL}}(t) \cdot \hat{X}\|_{CH(\rho_{m\gamma_j})}, \qquad s \in \mathbb{R},$$

where $\hat{X}$ is the Chow form for the algebraic cycle X sitting in $\mathbb{P}(V_{m\gamma_j}^*)$ by the Kodaira embedding via the complete linear system $|L^{\otimes m\gamma_j}|$. Then by Theorem 2.2,

$$\lim_{s\to-\infty} \dot{f}_{j,m}(s) = \frac{(n+1)! \, a_0}{|u_{j,m}|_1} (m\gamma_j)^{-n} \sum_{\alpha=1}^{\infty} \mathrm{DF}_\alpha(\mu_j) m^{n+1-\alpha}$$

$$= \frac{(n+1)! \, a_0}{|u_{j,m}|_1} \gamma_j^{-n} \sum_{\alpha=1}^{\infty} \mathrm{DF}_\alpha(\mu_j) m^{1-\alpha}, \qquad (5.18)$$

where $m \gg 1$ and $a_0 := \gamma_j^n c_1(L)^n[X]/n!$. Let $w_{j,m}$ be the weight of the $\mathbb{C}^*$-action on $\det \mathscr{E}_{j,m}$ via $\hat{\sigma}_{j,m}$. Put $N_{j,m} := \dim \mathscr{E}_{j,m}$. For $m \gg 1$, $w_{j,m}$ and $N_{j,m}$ are written as

$$
\begin{aligned}
w_{j,m} &= b_0 m^{n+1} + b_1 m^n + \cdots + b_n m + b_{n+1}, \\
N_{j,m} &= a_0 m^n + a_1 m^{n-1} + \cdots + a_n,
\end{aligned}
$$

so that dividing $w_{j,m}$ by $m N_{j,m}$, we obtain

$$
\frac{b_0 m^{n+1} + b_1 m^n + \cdots + b_n m + b_{n+1}}{a_0 m^{n+1} + a_1 m^n + \cdots + a_n m} = \frac{w_{j,m}}{m N_{j,m}}
$$

$$
= \sum_{\alpha=0}^{\infty} \mathrm{DF}_\alpha(\mu_j) m^{-\alpha} = \frac{b_0}{a_0} + \sum_{\alpha=1}^{\infty} \mathrm{DF}_\alpha(\mu_j) m^{-\alpha}. \tag{5.19}
$$

By (5.19), $\sum_{\alpha=1}^{\infty} \mathrm{DF}_\alpha(\mu_j) z^\alpha$ is a holomorphic function of z in a neighborhood of the origin with a zero at $z = 0$. Hence $\sum_{\alpha=1}^{\infty} \mathrm{DF}_\alpha(\mu_j) z^{\alpha-1}$ is holomorphic in z in a neighborhood of the origin. Hence, since $\|\mu_j\|_{asymp} = \lim_{m\to\infty} |u_{j,m}|_1$, it follows from (5.18) above that

$$
\lim_{m\to\infty} \left\{ \frac{\lim_{s\to-\infty} \dot{f}_{j,m}(s)}{(n+1)! \, a_0} \right\} = \frac{\mathrm{DF}_1(\mu_j)}{\|\mu_j\|_{asymp}} = \kappa(\mu_j). \tag{5.20}
$$

Take a sequence of positive real numbers ε_j, $j = 1, 2, \cdots$, such that $\varepsilon_j \to 0$ as $j \to \infty$. Then by (5.20), we may choose an increasing sequence of positive integers m_j, $j = 1, 2, \cdots$, satisfying both $m_j \gg 1$ and $\lim_{j\to\infty} m_j = +\infty$ such that

$$
\left| \frac{\lim_{s\to-\infty} \dot{f}_{j,m_j}(s)}{(n+1)! \, a_0} - \kappa(\mu_j) \right| \le \varepsilon_j, \qquad j = 1, 2, \cdots.
$$

In view of (2) above, by setting $\kappa_\infty := \lim_{j\to\infty} \kappa(\mu_j) \in \mathbb{R}_{\ge 0} \cup \{+\infty\}$, we have

$$
\lim_{s\to-\infty} \dot{f}_{j,m_j}(s) \to \kappa_\infty (n+1)! \, a_0, \quad \text{as } j \to \infty. \tag{5.21}
$$

By convexity of the function $f_{j,m_j}(s)$, its derivative $\dot{f}_{j,m_j}(s)$ is non-decreasing in s. Hence the following inequality holds for all $s \in \mathbb{R}$:

$$
\lim_{s\to-\infty} \dot{f}_{j,m_j}(s) \le \dot{f}_{j,m_j}(s). \tag{5.22}
$$

For special one-parameter groups $\sigma_{j,m_j} : \mathbb{C}^* \to \mathrm{GL}(V_{m\gamma_j})$, $j = 1, 2, \cdots$, by identifying the spaces $\mathscr{E}_{j,m_j}$ with $V_{m\gamma_j}$, we obtain test configurations

$$\mu_{j,m_j} = (\mathscr{X}_{\sigma_{j,m_j}}, \mathscr{L}_{\sigma_{j,m_j}}), \quad . j = 1, 2, \cdots .$$

Then the special linearization $\sigma_{j,m_j}^{\mathrm{SL}}$ is not necessarily an algebraic group homomorphism from $\mathbb{C}^*$ to $\mathrm{SL}(V_{m\gamma_j})$, but by choosing a suitable covering algebraic torus $\tilde{\mathbb{C}}^*$, the corresponding lift, denoted by $\tilde{\sigma}_j$, defines an algebraic group homomorphism

$$\tilde{\sigma}_j : \tilde{\mathbb{C}}^* \to \mathrm{SL}(V_{m\gamma_j}).$$

Then by setting $\tilde{\mu}_j := (\mathscr{X}_{\tilde{\sigma}_j}, \mathscr{L}_{\tilde{\sigma}_j})$, we see from Remark 4.5 that

$$F_1(\{\tilde{\mu}_j\}) = \lim_{s \to -\infty} \lim_{j \to \infty} \dot{f}_{j,m_j}(s).$$

It then follows from (5.21) and (5.22) that

$$0 \le \kappa_\infty (n+1)! a_0 = \lim_{j \to \infty} \left\{ \lim_{s \to -\infty} \dot{f}_{j,m_j}(s) \right\} \le \lim_{j \to \infty} \dot{f}_{j,m_j}(s), \qquad s \in \mathbb{R}.$$

Let $s \to -\infty$. Then $0 \le \lim_{s \to -\infty} \underline{\lim}_{j \to \infty} \dot{f}_{j,m_j}(s) = F_1(\{\tilde{\mu}_j\})$ in contradiction to the strong K-stability of (X, L), as required. $\square$

Theorem 5.5 (cf. [55]) *If (X, L) is strongly K-stable relative to T, then (X, L) is asymptotically Chow stable relative to T.*

Proof For each positive integer γ, let $\hat{X}_\gamma$ denote the Chow form for the image cycle of the twisted Kodaira embedding $X \subset \mathbb{P}(V_\gamma^*)$. Assume that (X, L) is strongly K-stable relative to T. By the notation in Sect. 4.3, it suffices to show the following:

(1) (X, L) is asymptotically Chow polystable relative to T.
(2) The isotropy subgroup of $T_\gamma^\perp$ at $\hat{X}_\gamma$ is finite for $\gamma \gg 1$.

In order to prove (1), we assume for contradiction that (X, L) is not asymptotically Chow polystable relative to T. Then there exists a sequence

$$1 \le \gamma_1 < \gamma_2 < \cdots < \gamma_j < \cdots$$

of integers γ_j such that $X \subset \mathbb{P}(V_{\gamma_j}^*)$ is not Chow polystable relative to T. Then by Corollary 5.1, there exists a $\sigma_j \in 1\mathrm{PS}(T_{\gamma_j}^\perp)$ for each j such that the Chow weight

$$\lambda_{\sigma_j} := \lim_{r \to -\infty} \frac{d\{\log \|\sigma_j(e^r) \cdot \hat{X}_j\|_{\mathrm{CH}(\rho_{\gamma_j})}\}}{dr} \qquad (5.23)$$

satisfies $\lambda_{\sigma_j} \geq 0$. Put $\hat{X}_j := \hat{X}_{\gamma_j}$ for simplicity. Consider the test configurations

$$\mu_j := (\mathscr{X}_{\sigma_j}, \mathscr{L}_{\sigma_j}), \qquad j = 1, 2, \cdots.$$

Let $u_j \neq 0$ be the fundamental generator for σ_j. Then for $t := e^r$, by setting $s = |u_j|_\infty r$, we obtain $t = \exp(s/|u_j|_\infty)$. Then for f_j in (4.4), we have

$$\dot{f}_j(s) = \frac{|u_j|_\infty}{|u_j|_1} \gamma_j^{-n} \frac{d\{\log \|\sigma_j(t) \cdot \hat{X}_j\|_{\mathrm{CH}(\rho_{\gamma_j})}\}}{dr} \cdot \frac{dr}{ds}, \qquad (5.24)$$

and let $s \to -\infty$. Then $r \to -\infty$, and by (5.23) and (5.24),

$$\lim_{s \to -\infty} \dot{f}_j(s) = \frac{|u_j|_\infty}{|u_j|_1} \gamma_j^{-n} \lambda_{\sigma_j} \frac{dr}{ds} = \frac{1}{|u_j|_1} \gamma_j^{-n} \lambda_{\sigma_j} \geq 0.$$

By the convexity of the function $f_j(s)$, its derivative $\dot{f}_j(s)$ is a non-decreasing function of s. Hence for each $s \in \mathbb{R}$,

$$0 \leq \lim_{s \to -\infty} \dot{f}_j(s) \leq \dot{f}_j(s).$$

In this inequality, let $j \to \infty$. Then for each $s \in \mathbb{R}$,

$$0 \leq \lim_{j \to \infty} \dot{f}_j(s).$$

We further let $s \to -\infty$. Then we obtain

$$F_1(\{\mu_j\}) = \lim_{s \to -\infty} \lim_{j \to \infty} \dot{f}_j(s) \geq 0,$$

in contradiction to the strong K-stability of the polarized algebraic manifold (X, L) relative to T. This finishes the proof of (1).

In order to prove (2), we assume for contradiction that there exists an increasing sequence of positive integers

$$1 \ll \gamma_1 < \gamma_2 < \cdots < \gamma_j < \cdots$$

such that the isotropy subgroup H_j of $T_{\gamma_j}^\perp$ at $\hat{X}_j$ satisfies $\dim H_j > 0$ for all j. By (1), the orbit $O_j := T_{\gamma_j}^\perp \cdot \hat{X}_j$ is closed in $W_j := W_{\gamma_j}$. Hence O_j is affine (and is a Stein space). Since $T_{\gamma_j}^\perp$ is reductive, a theorem of Matsushima [59] asserts that H_j is reductive. By the positivity of $\dim H_j$, the group H_j contains $\mathbb{C}^*$ as a subgroup, i.e.,

$$G_j (\cong \mathbb{C}^*) \subset H_j.$$

For a fixed maximal algebraic torus T' in $\text{Aut}^0(X)$ satisfying $T \subset T'$, let T'_c be the maximal compact subgroup of T'. We choose a reference metric $\omega = \text{Ric}(h)$ (by which we endow V_{γ_j} with the Hermitian structure ρ_{γ_j}) in such a way that both h and ω are invariant under the action of T'_c. Let P be the maximal connected linear algebraic subgroup of $\text{Aut}^0(X)$, and let P' be the identity component of the isotropy subgroup of $\text{SL}(V_{\gamma_j})$ at $[\hat{X}_j] \in \mathbb{P}(W_j)$. Since $0 \notin O_j$, the affine subset $O_j \cap \mathbb{C}\hat{X}_j$ of the affine line $\mathbb{C}\hat{X}_j$ is a finite set. Hence we have a natural isogeny

$$\iota : P' \to P.$$

Let T'_j be the identity component of $\iota^{-1}(T')$. Then T'_j is an algebraic torus containing T_{γ_j} as an algebraic subtorus. Note that Z_{γ_j} is the centralizer of T_{γ_j} in $\text{SL}(V_{\gamma_j})$. Clearly, both G_j and T'_j sit in the identity component, denoted by Q, of $Z_{\gamma_j} \cap P'$. Since T'_j is a maximal algebraic torus in in P', it is also a maximal algebraic torus in Q. Hence there exists an element g_0 in Q such that

$$g_0 \, G_j \, g_0^{-1} \subset T'_j. \tag{5.25}$$

Then by $g_0[\hat{X}_j] = [\hat{X}_j]$, there exists a $\lambda_0 \in \mathbb{C}^*$ such that $g_0 \hat{X}_j = \lambda_0 \hat{X}_j$, i.e., $\lambda_0^{-1}\hat{X}_j = g_0^{-1}\hat{X}_j$. If $g \in G_j$, then $g\hat{X}_j = \hat{X}_j$, so that

$$g_0 g g_0^{-1}\hat{X}_j = g_0 g(\lambda_0^{-1}\hat{X}_j) = \lambda_0^{-1}(g_0 g\hat{X}_j) = \lambda_0^{-1}(g_0\hat{X}_j) = \hat{X}_j.$$

Hence $g_0 G_j g_0^{-1}$ fixes $\hat{X}_j$. Let $0 \neq u$ be a basis for the one-dimensional Lie algebra $\mathfrak{g}_j := \text{Lie}(G_j)$. Then for all $v \in \mathfrak{t}_{\gamma_j}$,

$$\langle g_0 u g_0^{-1}, v\rangle_{\gamma_j} = \gamma_j^{-n-2} \text{Tr}((g_0 u g_0^{-1})v) = \gamma_j^{-n-2} \text{Tr}(u g_0^{-1} v g_0).$$

Now by $g_0 \in Q \subset Z_{\gamma_j}$ and $v \in \mathfrak{t}_{\gamma_j}$, we have $g_0^{-1} v g_0 = v$, so that the equality $g_0 u g_0^{-1} v = g_0 u v g_0^{-1}$ holds. Hence

$$\langle g_0 u g_0^{-1}, v\rangle_{\gamma_j} = \gamma_j^{-n-2} \text{Tr}(uv) = 0, \tag{5.26}$$

where the last equality follows from $u \in \mathfrak{t}_{\gamma_j}^{\perp}$. It then follows from (5.26) that

$$g_0 G_j g_0^{-1} \subset T_{\gamma_j}^{\perp}.$$

Since $g_0 G_j g_0^{-1}$ fixes $\hat{X}_j$, we obtain $g_0 G_j g_0^{-1} \subset H_j$. From this together with (5.25), we obtain a special one-parameter group $\sigma_j \in 1\text{PS}(T_{\gamma_j}^{\perp})$ such that

$$\sigma_j(\mathbb{C}^*) = g_0 G_j g_0^{-1} \subset H_j.$$

It then also follows that $\sigma_j^{-1} \in 1\mathrm{PS}(T_{\gamma_j}^{\perp})$ and $\sigma_j^{-1}(\mathbb{C}^*) \subset H_j$. We now consider the following sequences of test configurations:

$$\begin{aligned} \mu_j &= (\mathscr{X}_{\sigma_j}, \mathscr{L}_{\sigma_j}), & j &= 1, 2, \cdots, \\ \nu_j &= (\mathscr{X}_{\sigma_j^{-1}}, \mathscr{L}_{\sigma_j^{-1}}), & j &= 1, 2, \cdots. \end{aligned}$$

In view of (5.25), it is easily seen that

$$F_1(\{\mu_j\}) = -F_1(\{\nu_j\}). \tag{5.27}$$

On the other hand, since (X, L) is strongly K-stable relative to T, we have both

$$F_1(\{\mu_j\}) < 0 \quad \text{and} \quad F_1(\{\nu_j\}) < 0,$$

in contradiction to (5.27), as required. $\square$

Problems

5.1 Let $\mu := (\mathscr{X}, \mathscr{L})$ be a nontrivial normal test configuration, of exponent γ, for a polarized algebraic manifold (X, L). Consider the test configurations

$$\mu_j = (\mathscr{X}, \mathscr{L}^{\otimes j}), \qquad j = 1, 2, \cdots, \tag{5.28}$$

of exponent $\gamma_j := j\gamma$, for (X, L). Then by Sect. 2.2, each μ_j is a test configuration μ_{σ_j} associated to a special one-parameter group $\sigma_j : \mathbb{C}^* \to \mathrm{GL}(V_{\gamma_j})$. Hence we have

$$F_1(\{\mu_j\}) \in \mathbb{R} \cup \{-\infty\}$$

as in Remark 4.5. Assume that the inequality

$$F_1(\{\mu_j\}) \geq \frac{(n+1)c_1(L)^n[X]}{\|\mu\|_{asymp}} \mathrm{DF}_1(\mu) \tag{5.29}$$

always holds. Show that, if (X, L) is strongly K-stable, then (X, L) is K-stable.

5.2 Let μ_j, $j = 1, 2, \cdots$, be a sequence of test configurations μ_{σ_j} of exponent γ_j for (X, L) associated to special one-parameter groups

$$\sigma_j : \mathbb{C}^* \to \mathrm{SL}(V_{\gamma_j}), \qquad j = 1, 2, \cdots.$$

such that $F_1(\{\mu_j\}) = 0$ (cf. Remark 4.6). Assuming that (X, L) is strongly K-stable, show that there exists a positive integer N such that σ_j are trivial for all $j \geq N$.

Chapter 6
The Yau–Tian–Donaldson Conjecture

Abstract In this chapter, we discuss the Yau–Tian–Donaldson conjecture from a historical point of view.

- In Sect. 6.1, we briefly discuss the Calabi conjecture. The unsolved case of the Calabi conjecture motivates the Yau–Tian–Donaldson conjecture in Kähler–Einstein cases.
- As mentioned in Sect. 6.2, the Yau–Tian–Donaldson conjecture in Kähler–Einstein cases was solved affirmatively by Chen, Donaldson and Sun and by Tian.
- In Sect. 6.3, we define K-energy and modified K-energy for compact Kähler manifolds. This concept allows us to state the recent results of Chen and Cheng and of He on the existence of CSC Kähler metrics and extremal Kähler metrics.
- Finally, in Sect. 6.4, various versions of the Yau–Tian–Donaldson conjecture will be considered in extremal Kähler cases.

Keywords The Calabi conjecture · The Yau–Tian–Donaldson conjecture · The modified K-energy

6.1 The Calabi Conjecture

For a compact complex connected manifold X, we view $c_1(X)$ as a Dolbeault cohomology class on X. The following conjecture posed by Calabi [6, 7] in the 1950s on the existence of Kähler–Einstein metrics is the *Calabi conjecture*:

1. If $c_1(X) < 0$, then X admits a unique Kähler–Einstein form ω in the canonical class $-c_1(X)$ such that $\text{Ric}(\omega) = -\omega$.
2. If $c_1(X) = 0$, then for every Kähler class $\mathscr{K}$ on X, there exists a unique Kähler–Einstein form $\omega \in \mathscr{K}$ such that $\text{Ric}(\omega) = 0$.
3. More generally, for every d-closed $(1, 1)$-form η in the class $c_1(X)$, every Kähler class $\mathscr{K}$ on X admits a unique Kähler form ω such that $\text{Ric}(\omega) = \eta$.

It is well-known that the Calabi conjecture was solved affirmatively in the 1970s. Yau [86, 87] gave an affirmative answer to the whole conjecture by solving complex

T. Mabuchi, *Test Configurations, Stabilities and Canonical Kähler Metrics*,
SpringerBriefs in Mathematics, https://doi.org/10.1007/978-981-16-0500-0_6

Monge–Ampère equations using the continuity method, whereas (1) was proved independently also by Aubin [1]. These works represent a landmark in differential geometric studies of algebraic geometry.

In the case where $c_1(X) > 0$, as advocated by Yau [88], new developments are made from the viewpoints of stability of the underlying manifolds. Related to the existence problem of Kähler–Einstein metrics on Fano manifolds, Tian [79] introduced the concept of K-stability, while Donaldson [20] reformulated the concept in a general setting by purely algebraic geometric languages.

6.2 The Yau–Tian–Donaldson Conjecture

A compact complex connected manifold X is called *Fano* if its first Chern class $c_1(X)$ is positive. As to the existence of Kähler–Einstein metrics on Fano manifolds, the following is known as the *Yau–Tian–Donaldson conjecture*:

Conjecture 6.1 A Fano manifold X admits a Kähler–Einstein metric if and only if the polarized algebraic manifold (X, K_X^{-1}) is K-polystable.

Recently, this conjecture was solved affirmatively by Chen et al. [13–15] and Tian [80] through the use of cone singularities applying the Cheeger–Colding–Tian theory. However, for CSC Kähler metrics and extremal Kähler metrics, the problem of characterizing the existence of such metrics by stability of underlying manifolds is still open in spite of various partial results.

6.3 The K-Energy

Let $\mathscr{K}$ be a Kähler class on a compact complex connected manifold X. Fix a reference Kähler form $\omega_0 \in \mathscr{K}$. Then φ_t, $0 \leq t \leq 1$, is called a *piecewise smooth path* in $C^\infty(X)_{\mathbb{R}}$ if the mapping

$$[0, 1] \times X \to \mathbb{R}, \qquad (t, x) \mapsto \varphi_t(x),$$

is continuous and moreover there exists a partition $0 = a_0 < a_1 < \cdots < a_r = 1$ of the unit interval $[0, 1]$ such that the restrictions

$$[a_{i-1}, a_i] \times X \to \mathbb{R}, \qquad (t, x) \mapsto \varphi_t(x),$$

are C^∞, $i = 1, 2, \cdots, r$. Put $V := \int_X \omega^n$. Every ω in $\mathscr{K}$ is expressible as ω_φ for some $\varphi \in C^\infty(X)_{\mathbb{R}}$, where ω_φ is the Kähler form in the class $\mathscr{K}$ defined by

$$\omega_\varphi := \omega_0 + dd^c\varphi = \omega_0 + \frac{\sqrt{-1}}{2\pi}\partial\bar{\partial}\varphi.$$

Let φ_t, $0 \leq t \leq 1$, be a piecewise smooth path in $C^\infty(X)_\mathbb{R}$ satisfying $\varphi_0 = 0$, $\varphi_1 = \varphi$ and $\omega_{\varphi_t} \in \mathcal{K}$. Put $\dot{\varphi}_t = \partial\varphi_t/\partial t$. We then define the *K-energy* $\kappa : \mathcal{K} \to \mathbb{R}$ for $\mathcal{K}$ by

$$\kappa(\omega) := -\frac{1}{V} \int_0^1 \left\{ \int_X \dot{\varphi}_t \, (S(\omega_{\varphi_t}) - S_0) \, \omega_{\varphi_t}^n \right\} dt, \tag{6.1}$$

where $S(\omega_{\varphi_t})$ is the scalar curvature of ω_{φ_t}, and S_0 is the average of the scalar curvature of ω_0 defined by $S_0 := \int_X S(\omega_0)\omega_0^n / \int_X \omega_0^n$. Then (cf. [42]):

Lemma 6.1 *The right-hand side of* (6.1) *depends only on* ω, *and is independent of the choice of the path* $\{\varphi_t\}_{0 \leq t \leq 1}$.

Proof Define $\psi = \psi(s, t)$ by setting $\psi := s\varphi_t$ for $(s, t) \in [0, 1] \times [0, 1]$. Let $\Psi = \Psi(s, t)$ be the 1-form on $[0, 1] \times [0, 1]$ defined by

$$\Psi := -\left\{ \int_X \psi_s(S(\omega_\psi) - S_0)\omega_\psi^n \right\} ds - \left\{ \int_X \psi_t(S(\omega_\psi) - S_0)\omega_\psi^n \right\} dt$$

$$= -n \int_X (\psi_s \, ds + \psi_t \, dt)\{\text{Ric}(\omega_\psi) - \lambda\omega_\psi\} \wedge \omega_\psi^{n-1},$$

where $\psi_s := \frac{\partial\psi}{\partial s}$, $\psi_t := \frac{\partial\psi}{\partial t}$ and $\lambda := S_0/n$. Then by $\text{Ric}(\omega_\psi) = -dd^c \log \omega_\psi^n$, we have $(\partial/\partial s)\,\text{Ric}(\omega_\psi) = -dd^c(\Delta_\psi\psi_s)$ and $(\partial/\partial t)\,\text{Ric}(\omega_\psi) = -dd^c(\Delta_\psi\psi_t)$. Hence

$$d\Psi = -n \int_X (\psi_s \, ds + \psi_t \, dt)\left(dd^c\{(\Delta_\psi + \lambda)\psi_s\}ds + dd^c\{(\Delta_\psi + \lambda)\psi_t\}dt \right) \wedge \omega_\psi^{n-1}$$

$$- n \int_X (\psi_s \, ds + \psi_t \, dt)\{\text{Ric}(\omega_\psi) - \lambda\omega_\psi\} \wedge \theta,$$

where $\theta := (n-1)\omega_\psi^{n-2}\{ds(dd^c\psi_s) + dt(dd^c\psi_t)\}$. The second integral on the right-hand side is obviously zero, while the first integral is

$$-ds \wedge dt \int_X \left\{ \psi_s \, \Delta_\psi(\Delta_\psi + \lambda)\psi_t - \psi_t \, \Delta_\psi(\Delta_\psi + \lambda)\psi_s \right\} \omega_\psi^n = 0.$$

Hence $d\Psi = 0$. For the 2-chain $\square = [0, 1] \times [0, 1]$ in the plane $\mathbb{R}^2 = \{(s, t)\}$, we consider its boundary $\partial\square = \sigma_1 + \sigma_2 - \sigma_3 - \sigma_4$, where $\sigma_1 = \{(s, 0); 0 \leq s \leq 1\}$, $\sigma_2 = \{(1, t); 0 \leq t \leq 1\}$, $\sigma_3 = \{(s, 1); 0 \leq s \leq 1\}$, $\sigma_4 = \{(0, t); 0 \leq t \leq 1\}$. It now follows from $d\Psi = 0$ and $\int_{\sigma_4} \Psi = 0$ that

$$0 = \int_\square d\Psi = \int_{\partial\square} \Psi = \int_{\sigma_1 + \sigma_2 - \sigma_3 - \sigma_4} \Psi = \int_{\sigma_2} \Psi - \int_{\sigma_3 - \sigma_1} \Psi. \tag{6.2}$$

Since $\varphi_0 = 0$, we have $\int_{\sigma_1} \Psi = 0$. Hence by (6.2), $\int_{\sigma_2} \Psi = \int_{\sigma_3} \Psi$, i.e.,

$$-\frac{1}{V} \int_0^1 \left\{ \int_X \dot{\varphi}_t (S(\omega_{\varphi_t}) - S_0) \omega_{\varphi_t}^n \right\} dt = -\frac{1}{V} \int_0^1 \left\{ \int_X \varphi \{ S(\omega_{s\varphi}) - S_0 \} \omega_{s\varphi}^n \right\} ds.$$

Thus, (6.1) is independent of the choice of the path $\{\varphi_t\}_{0 \le t \le 1}$, as required. □

For $\omega = \omega_\varphi$ above, $\kappa(\omega)$ is written also as $\kappa(\varphi)$. By setting $\omega_t := \omega_{\varphi_t}$ in (6.1), we now define the modified K-energy ([28, 73]; see also [46]) as follows:

$$\hat{\kappa}(\varphi) := -\frac{1}{V} \int_0^1 \left\{ \int_X \dot{\varphi}_t \left(S(\omega_t) - S_0 - H_{\omega_t} \right) \omega_t^n \right\} dt,$$

where the Hamiltonian function H_{ω_t} on (X, ω_t) associated to the extremal vector field y_{ω_t} as in Sect. 9.2 is defined by

$$\mathrm{grad}_{\omega_t}^{\mathbb{C}} H_{\omega_t} = y_{\omega_t} = y_{\omega_0} \quad \text{and} \quad \int_X H_{\omega_t} \omega_t^n = 0.$$

Here the condition $y_{\omega_t} = y_{\omega_0}$ corresponds to the condition that φ_t is a function invariant under the S^1-action generated by the real vector field $y_{\omega_0}^{\mathbb{R}} := y_{\omega_0} + \bar{y}_{\omega_0}$.

By recent works of Chen and Cheng [10–12] and He [29], the existence of CSC Kähler metrics and extremal Kähler metrics is characterized by the properness of the modified K-energy modulo a suitable subgroup of $\mathrm{Aut}^0(X)$ (see Sect. 8.1 in CSC Kähler cases; see also Theorem 9.7, Sect. 9.6, in extremal Kähler cases).

6.4 Extremal Kähler Versions of the Conjecture

For a polarized algebraic manifold (X, L), by choosing a maximal compact subgroup K of $\mathrm{Aut}^0(X)$, we view its complexification $K^{\mathbb{C}}$ as a connected reductive algebraic subgroup of $\mathrm{Aut}^0(X)$. Let Z be the central algebraic torus in $K^{\mathbb{C}}$, and we also consider a maximal algebraic torus $T_{\max}$ in $K^{\mathbb{C}}$. Let T be either $T_{\max}$ or Z.

Existence Problem of Extremal Kähler Metrics *Find a necessary and sufficient stability condition for X to admit an extremal Kähler metric in the class $c_1(L)$.*

As an extremal Kähler version of the Yau–Tian–Donaldson conjecture, we expect that the required stability condition is one of the following:

K-polystability, uniform K-stability, strong K-stability

relative to either $T_{\max}$ or Z. It could also occur that these stability conditions partially coincide. Existence and stability theorems for extremal Kähler metrics will be discussed in the next two chapters.

Problems

6.1 For a polarized algebraic manifold (X, L), let T and T' be algebraic tori in $\text{Aut}^0(X)$ such that $T \subset T'$. Show that, if (X, L) is uniformly K-stable relative to T, then (X, L) is uniformly K-stable relative to T'.

6.2 For (X, L) and $T_{\max}$ in Sect. 6.4, let $\mathscr{E}_{\text{EX}}$ be the set of all extremal Kähler metrics in the class $c_1(L)$, and $\mathscr{E}_{\text{CSC}}$ the set of all constant scalar curvature Kähler metrics in the class $c_1(L)$. Consider the following conjectures:

$$\mathscr{E}_{\text{EX}} \neq \emptyset \iff (X, L) \text{ is uniformly K-stable relative to } T_{\max}, \tag{1}$$

$$\mathscr{E}_{\text{CSC}} \neq \emptyset \iff \mathscr{F} = 0 \text{ and } (X, L) \text{ is uniformly K-stable relative to } T_{\max}, \tag{2}$$

where $\mathscr{F}$ is the Futaki character as in Sect. 3.1. Show that, if Conjecture (1) is true, then Conjecture (2) is true.

Chapter 7
Stability Theorem

Abstract In this chapter, the existence of CSC Kähler metrics or extremal Kähler metrics on a polarized algebraic manifold implies various kinds of stability.

- In Sect. 7.1, we shall show that the existence of a CSC Kähler metric implies strong K-semistability of the polarized algebraic manifolds.
- In Sect. 7.2, we introduce the concept of relative balanced metrics. The existence of such a metric corresponds to relative Chow polystability.
- In Sect. 7.3, we shall show that the existence of an extremal Kähler metric implies strong K-semistability of the polarized algebraic manifolds.
- In Sect. 7.4, a result of Stoppa and Székelyhidi asserts that a polarized algebraic manifold (X, L) with an extremal Kähler metric in $c_1(L)$ is K-polystable relative to a maximal algebraic torus $T_{\max}$ in $\mathrm{Aut}^0(X)$.
- Finally, in Sect. 7.5, we have an inequality by switching the order of the double limit in the definition of the invariant $F_1(\{\mu_j\})$.

Keywords Stability for extremal Kähler manifolds · Relative balanced metrics

7.1 Strong K-Semistability of CSC Kähler Manifolds

For a polarized algebraic manifold (X, L), assume that the Kähler class $c_1(L)$ admits a CSC Kähler form ω. Let h be a Hermitian metric on L such that $\mathrm{Ric}(h) = \omega$. We then consider a sequence $\{\mu_j\}$ of test configurations for (X, L) such that

$$\mu_j = (\mathscr{X}_{\sigma_j}, \mathscr{L}_{\sigma_j}), \qquad j = 1, 2, \cdots,$$

where $\sigma_j : \mathbb{C}^* \to \mathrm{SL}(V_{\gamma_j})$, $j = 1, 2, \cdots$, are nontrivial special one-parameter groups satisfying the condition

$$\gamma_j \to +\infty, \quad \text{as } j \to \infty.$$

© The Author(s), under exclusive licence to Springer Nature Singapore Pte Ltd. 2021
T. Mabuchi, *Test Configurations, Stabilities and Canonical Kähler Metrics*,
SpringerBriefs in Mathematics, https://doi.org/10.1007/978-981-16-0500-0_7

Let u_j be the fundamental generator of σ_j. We then consider the functions

$$f_j(s) = \frac{|u_j|_\infty}{|u_j|_1} \gamma_j^{-n} \log \|\sigma_j(t) \cdot \hat{X}_j\|_{CH(\rho_{\gamma_j})}, \qquad j = 1, 2, \cdots,$$

as in (4.4), where $t := \exp(s/|u_j|_\infty)$ for $s \in \mathbb{R}$. As in the proof of Claim in Sect. 4.5, let $-\beta_k$, $k = 1, 2, \cdots, N_j$, be the weights of the $\mathbb{C}^*$-action on $(V_{\gamma_j}, \rho_{\gamma_j})$, and for its suitable orthonormal basis $\{v_1, \cdots, v_{N_j}\}$, we can write

$$\sigma_j(t) \cdot v_k = t^{-\beta_k} v_k, \qquad k = 1, 2, \cdots, N_j,$$

such that $\beta_1 + \cdots + \beta_{N_j} = 0$. Here by (4.6),

$$B_j(\omega) = C_j + O(\gamma_j^{-2}), \tag{7.1}$$

where for $j \gg 1$, we have $C_j := 1 + (1/2)S_\omega\gamma_j^{-1} > 0$. In our case, since the extremal vector field y $(= y_\gamma)$ is zero, $B_j^\#(\omega)$ in (4.7) is just $B_j(\omega)$, and we adapt the proof of Claim in Sect. 4.5 to our case by setting $v_k^\# = v_k$ for all k. Then by (4.8) and (7.1),

$$\dot{f}_j(0) = \frac{(n+1)!}{|u_j|_1} \gamma_j^{-n} \int_X \frac{\sum_{\alpha=1}^{N_j} \beta_k |v_k|_h^2}{C_j + O(\gamma_j^{-2})} \{\omega + O(\gamma_j^{-2})\}^n$$

$$= \frac{(n+1)!}{C_j|u_j|_1} \gamma_j^{-n} \int_X (\sum_{\alpha=1}^{N_j} \beta_k |v_k|_h^2)\{\omega^n + O(\gamma_j^{-2})\},$$

where the right-hand side just above is

$$\frac{(n+1)!}{C_j|u_j|_1} \gamma_j^{-n} \left\{ \sum_{k=1}^{N_j} \beta_k + \sum_{k=1}^{N_j} |\beta_k| O(\gamma_j^{-2}) \right\} = \frac{O(\sum_{k=1}^{N_j} |\beta_k|)}{\gamma_j^{n+2}|u_j|_1} = O(\gamma_j^{-1}).$$

Hence by $\lim_{j \to \infty} \gamma_j^{-1} = 0$, we have $\underline{\lim}_{j \to \infty} \dot{f}_j(0) = 0$. Since $\underline{\lim}_{j \to \infty} \dot{f}_j(s)$ is a non-decreasing function of s, we finally obtain

$$F_1(\{\mu_j\}) = \lim_{s \to -\infty} \underline{\lim}_{j \to \infty} \dot{f}(s) \leq 0,$$

i.e., (X, L) is strongly K-semistable.

7.2 Relative Balanced Metrics

For a polarized algebraic manifold (X, L), let T be an algebraic torus in $\mathrm{Aut}^0(X)$. Let γ be a positive integer. By using the notation in Sect. 4.3, let $(\mathfrak{t}_\gamma)_c$ denote the Lie algebra of the maximal compact subgroup $(T_\gamma)_c$ of T_γ. Put $(\mathfrak{t}_\gamma)_{\mathbb{R}} := \sqrt{-1}(\mathfrak{t}_\gamma)_c$. Recall that the space $V_\gamma = H^0(X, L^{\otimes \gamma})$ is written as a direct sum (see Sect. 4.3)

$$V_\gamma = \bigoplus_{i=1}^{n_\gamma} V_{\gamma, i},$$

where $V_{\gamma, i} = \{v \in V_\gamma ; \theta \sigma = \chi_{\gamma, i}(\theta) \sigma \text{ for all } \theta \in \mathfrak{t}_\gamma\}$. For a Kähler metric ω in the class $c_1(L)$, we choose a Hermitian metric h for L such that $\mathrm{Ric}(h) = \omega$. As in (2.2), we have a natural Hermitian structure ρ_γ on V_γ.

The concept of balanced metrics introduced in Chap. 5 plays a very important role in the study of CSC Kähler metrics. A similar concept exists in the study of extremal Kähler metrics. Actually, we have a relative version of balanced metrics:

Definition 7.1 ω is called a *balanced metric for $L^{\otimes \gamma}$ relative to T* if there exist a $\theta_\gamma \in (\mathfrak{t}_\gamma)_{\mathbb{R}}$ and a positive constant C such that $1 - \chi_{\gamma, i}(\theta_\gamma) > 0$ for all i and that

$$\sum_{i=1}^{n_\gamma} \sum_{\alpha=1}^{q_i} (1 - \chi_{\gamma, i}(\theta_\gamma)) |v_{i,\alpha}|_h^2 = C,$$

where $\{v_{i,\alpha} ; \alpha = 1, 2, \cdots, q_i, i = 1, 2, \cdots n_\gamma\}$ is an orthonormal basis for (V_γ, ρ_γ) such that each $\{v_{i,\alpha}; \alpha = 1, \cdots, q_i\}, i = 1, \cdots, n_\gamma$, is a basis for $V_{\gamma, i}$.

As in Sect. 5.1, let $X \subset \mathbb{P}(V_\gamma^*)$ be an abstractly defined Kodaira embedding associated to the complete linear system $|L^{\otimes \gamma}|$ on X. As a relative version of Theorem 5.1, we have the following facts (see for instance [48, 50, 54] for more details):

Fact 1 $X \subset \mathbb{P}(V_\gamma^*)$ is Chow polystable relative to T if and only if there exists a balanced metric ω for $L^{\otimes \gamma}$ relative to T.

Fact 2 $\chi_{\gamma, i}(\theta_\gamma) = \gamma^{-2} \chi_{\gamma, i}(\sqrt{-1} y_\gamma) + O(\gamma^{-2})$, and in Sect. 4.4, $|v_{i,\alpha}^\#|^2$ is written as $\{1 - \chi_{\gamma, i}(\theta_\gamma) + O(\gamma^{-2})\} |v_{i,\alpha}|^2$, where $O(\gamma^{-2})$ is a quantity whose absolute value is bounded from above by $C\gamma^{-2}$ for some positive constant C independent of i, γ, α.

Put $(y_\gamma)_{\mathbb{R}} := \sqrt{-1} y_\gamma$ and $b_{i,\alpha} := \gamma^{-2} \chi_{\gamma, i}(\sqrt{-1} y_\gamma) = O(\gamma^{-1})$. Then by the basis $\{v_{i,\alpha} ; \alpha = 1, 2, \cdots, q_i, i = 1, 2, \cdots, n_\gamma\}$ for V_γ, we can write $\gamma^{-2}(y_\gamma)_{\mathbb{R}} \in \mathfrak{t}_\gamma$ as a diagonal matrix with the (i, α)-th component equal to $b_{i,\alpha}$. Note also that

$$|v_{i,\alpha}^\#|^2 = \{1 - \gamma^{-2} \chi_{\gamma, i}(\sqrt{-1} y_\gamma)\} |v_{i,\alpha}|^2 = (1 - b_{i,\alpha}) |v_{i,\alpha}|^2. \tag{7.2}$$

7.3 Strong Relative K-Semistability of Extremal Kähler Manifolds

In this section, for a polarized algebraic manifold (X, L), we consider the situation that X admits an extremal Kähler metric in the class $c_1(L)$. Then by a result of Calabi [9], the centralizer G_0 in $\mathrm{Aut}(X)$ of an extremal vector field (cf. Sect. 9.2) is a reductive algebraic group. For the center $Z(G_0)$ of G_0, we choose an arbitrary algebraic torus T satisfying

$$Z(G_0) \subset T \subset G_0,$$

where S^1 generated by the extremal vector field, being a subgroup of the maximal compact subgroup of the center $Z(G_0)$, sits also in the maximal compact subgroup of T. The following generalization of a result of Donaldson [19] is known:

Fact (cf. [54, 72]; see also [71]) *If X admits an extremal Kähler metric in the class $c_1(L)$, then (X, L) is asymptotically Chow stable relative to T.*

Let ω be an extremal Kähler metric in the class $c_1(L)$. Choose a Hermitian metric h for L such that $\mathrm{Ric}(h) = \omega$. Then we also have the following:

Theorem 7.1 *If X admits an extremal Kähler metric ω in the Kähler class $c_1(L)$, then (X, L) is strongly K-semistable relative to T above.*

Proof For an extremal Kähler metric ω in the class $c_1(L)$, let $y := \mathrm{grad}_\omega^{\mathbb{C}} S_\omega$ be the associated extremal vector field. By using the notation in Sects. 4.4 and 4.5, we consider the twisted Bergman kernel

$$B_\gamma^\#(\omega) := (n!/\gamma^n) \sum_{k=1}^{N_\gamma} |v_k^\#|_h^2.$$

Put $r_0 := \{2c_1(L)^n[X]\}^{-1}\{nc_1(L)^{n-1}c_1(X)[X] + \sqrt{-1}\int_X h^{-1}(yh)\omega^n\}$. Then by [48] (see also [50, 51, 54]), we obtain

$$B_\gamma^\#(\omega) = 1 + r_0\gamma^{-1} + O(\gamma^{-2}). \tag{7.3}$$

Let $\{\mu_j\} \in \mathcal{M}_T$, i.e., $\mu_j = (\mathcal{X}_{\sigma_j}, \mathcal{L}_{\sigma_j})$ for some $\sigma_j \in 1\mathrm{PS}(T_{\gamma_j}^\perp)$ such that $\gamma_j \to +\infty$ as $j \to \infty$. Let $0 \neq u_j \in \mathfrak{t}_{\gamma_j}^\perp$ be the fundamental generator for σ_j. By setting $t := \exp(s/|u_j|_\infty)$, $s \in \mathbb{R}$, we consider a real-valued function $f_j(s)$ on $\mathbb{R}$ defined by (4.4). Then for a suitable choice of an admissible basis $\{v_1^\#, \cdots, v_{N_j}^\#\}$ for V_{γ_j}, we can write

$$\sigma_j(t) \cdot v_k^\# = t^{-\beta_k}v_k^\#, \qquad k = 1, 2, \cdots, N_j,$$

where $-\beta_1, \cdots, -\beta_{N_j}$ are the weights of the $\mathbb{C}^*$-action on V_{γ_j} via σ_j satisfying the equality $\beta_1 + \cdots + \beta_{N_j} = 0$. By taking $dd^c \log$ of both sides of (7.3), we obtain

$$\omega_{\mathrm{FS}} - \gamma_j \omega = O(\gamma^{-2}), \qquad (7.4)$$

where $\omega_{\mathrm{FS}} := dd^c \log \sum_{k=1}^{N_\gamma} |v_k^\#|^2$. In view of (7.3) and (7.4),

$$\dot{f}_j(0) = \frac{1}{|u_j|_1} \gamma_j^{-n}(n+1) \int_X \frac{\beta_1 |v_1^\#|_h^2 + \cdots + \beta_{N_j} |v_{N_j}^\#|_h^2}{|v_1^\#|_h^2 + \cdots + |v_{N_j}^\#|_h^2} \, \omega_{\mathrm{FS}}^n$$

$$= \frac{1}{|u_j|_1} \gamma_j^{-n}(n+1)! \int_X \frac{\beta_1 |v_1^\#|_h^2 + \cdots + \beta_{N_j} |v_{N_j}^\#|_h^2}{B_{\gamma_j}^\#(\omega)} \{\omega + O(\gamma_j^{-3})\}^n$$

$$= \frac{\gamma_j^{-n}(n+1)!}{|u_j|_1} \int_X \frac{\beta_1 |v_1^\#|_h^2 + \cdots + \beta_{N_j} |v_{N_j}^\#|_h^2}{1 + r_0 \gamma_j^{-1} + O(\gamma_j^{-2})} \{\omega + O(\gamma_j^{-3})\}^n.$$

By renumbering $\{v_k^\#;\ k = 1, 2, \ldots, N_j\}$, we obtain $\{v_{i,\alpha}^\#\}$ as in Sect. 4.4. Similarly, the weights β_k, $k = 1, 2, \ldots, N_j$, are renumbered as $\beta_{i,\alpha}$, $\alpha = 1, 2, \ldots, q_i$, $i = 1, 2, \ldots, n_\gamma$. Then by (7.2),

$$\beta_1 |v_1^\#|_h^2 + \cdots + \beta_{N_j} |v_{N_j}^\#|_h^2 = \sum_{i=1}^{n_\gamma} \sum_{\alpha=1}^{q_i} \beta_{i,\alpha}(1 - b_{i,\alpha}) |v_{i,\alpha}|_h^2.$$

Hence, in view of the fact that $\{v_{i,\alpha};\ \alpha = 1, 2, \ldots, q_i, i = 1, 2, \ldots, n_\gamma\}$ is an orthonormal basis for $(V_{\gamma_j}, \rho_{\gamma_j})$ and that $b_{i,\alpha} = O(\gamma_j^{-1})$, we obtain

$$\dot{f}_j(0) = \frac{\gamma_j^{-n}(n+1)!}{|u_j|_1} \int_X \frac{\sum_{i=1}^{n_\gamma} \sum_{\alpha=1}^{q_i} \beta_{i,\alpha}(1 - b_{i,\alpha}) |v_{i,\alpha}|_h^2}{1 + r_0 \gamma_j^{-1} + O(\gamma_j^{-2})} \{\omega + O(\gamma_j^{-3})\}^n$$

$$= \frac{\gamma_j^{-n}(n+1)!}{|u_j|_1} \left\{ \frac{\sum_{i=1}^{n_\gamma} \sum_{\alpha=1}^{q_i} \beta_{i,\alpha}(1 - b_{i,\alpha})}{1 + r_0 \gamma_j^{-1}} + \sum_{i=1}^{n_\gamma} \sum_{\alpha=1}^{q_i} |\beta_{i,\alpha}| \, O(\gamma_j^{-2}) \right\}.$$

Since $u_j \in \mathfrak{t}_{\gamma_j}^\perp$ is a diagonal matrix with (i, α)-th diagonal element $-\beta_{i,\alpha}$, and since $\gamma_j^{-2}(y_{\gamma_j})_{\mathbb{R}} \in (\mathfrak{t}_{\gamma_j})_{\mathbb{R}}$ is a diagonal matrix with (i, α)-th diagonal element $b_{i,\alpha}$, we see from the perpendicularity $\mathfrak{t}_{\gamma_j}^\perp \perp \mathfrak{t}_{\gamma_j}$ that $\sum_{i=1}^{n_\gamma} \sum_{\alpha=1}^{q_i} \beta_{i,\alpha} b_{i,\alpha} = 0$. We also have $\sum_{i=1}^{n_\gamma} \sum_{\alpha=1}^{q_i} \beta_{i,\alpha} = \beta_1 + \cdots + \beta_{N_j} = 0$. Hence

$$\dot{f}_j(0) = O(\gamma_j^{-1}) \cdot \frac{\sum_{i=1}^{n_\gamma} \sum_{\alpha=1}^{q_i} |\beta_{i,\alpha}|}{\gamma_j^{n+1}} \cdot \frac{1}{|u_j|_1} = O(\gamma_j^{-1}).$$

In view of the condition $\gamma_j \to +\infty$ as $j \to \infty$, we see that $\varliminf_{j\to\infty} \dot{f}_j(0) = 0$. Since $\varliminf_{j\to\infty} \dot{f}_j(s)$ is a non-decreasing function of s, it follows that

$$F_1(\{\mu_j\}) = \lim_{s\to-\infty} \varliminf_{j\to\infty} \dot{f}_j(s) \le \varliminf_{j\to\infty} \dot{f}_j(0) = 0.$$

Hence (X, L) is strongly K-semistable, as required.   □

7.4   K-Polystability of Extremal Kähler Manifolds

For a polarized algebraic manifold (X, L), let $T_{\max}$ be a maximal algebraic torus in $\mathrm{Aut}^0(X)$. Then the following theorem of Stoppa and Székelyhidi [75] holds:

Theorem 7.2   *If X admits an extremal Kähler metric in the class $c_1(L)$, then (X, L) is K-polystable relative to $T = T_{\max}$.*

As the algebraic torus T gets smaller, the condition of relative K-polystability gets stronger. Hence, as far as the stability of extremal Kähler manifolds is concerned, we want to choose T as small as possible.

7.5   A Reformulation of the Definition of the Invariant $F(\{\mu_j\})$

Let $\mu_j := (\mathscr{X}_{\sigma_j}, \mathscr{L}_{\sigma_j})$, $j = 1, 2, \cdots$, be a sequence of test configurations for (X, L) associated to nontrivial special one-parameter groups $\sigma_j : \mathbb{C}^* \to \mathrm{SL}(V_{\gamma_j})$, where $\gamma_j \to +\infty$ as $j \to \infty$. From now on, $(\mathscr{X}_{\sigma_j}, \mathscr{L}_{\sigma_j})$ will be written simply as $(\mathscr{X}_j, \mathscr{L}_j)$. By setting $t = \exp(s/|u_j|_\infty)$, we consider the function

$$f_j(s) := \frac{|u_j|_\infty}{|u_j|_1} \gamma_j^{-n} \log \|\sigma_j(t) \cdot \hat{X}_j\|_{\mathrm{CH}(\rho_{\gamma_j})}, \qquad s \in \mathbb{C},$$

where u_j is the fundamental generator of σ_j^{SL}, and $0 \ne \hat{X}_j \in W_j := \mathrm{Sym}^{d_j}(V_{\gamma_j}^*)^{\otimes n+1}$ is the Chow form as in (4.4) for the image X_j of the twisted Kodaira embedding. Let

$$\mathscr{N} := \langle \mathscr{L}_j, \mathscr{L}_j, \cdots, \mathscr{L}_j \rangle_{\mathscr{X}_j/\mathbb{A}^1}$$

be the Deligne pairing of $(n + 1)$-pieces of $\mathscr{L}_j$. Fix a Hermitian metric h for L such that $\omega := \mathrm{Ric}(h)$ is Kähler. For a suitable orthonormal basis $\{v_1, \cdots, v_{N_j}\}$ for V_{γ_j}, we view $\mathbb{P}(V_{\gamma_j}^*)$ as the projective space $\mathbb{P}^{N_j-1}(\mathbb{C}) := \{(z_1 : z_2 : \cdots : z_{N_j})\}$. Since

$\mathscr{X}_j \subset \mathbb{P}(V_{\gamma_j}^*) \times \mathbb{A}^1$, by viewing $(\mathscr{X}_j)_1$ as X, we have the identification

$$(\mathscr{X}_j)_t = \sigma_j(t) \cdot X \subset \mathbb{P}^{N_j-1}(\mathbb{C}), \qquad t \in \mathbb{C}^*. \qquad (7.5)$$

Let pr_1 be the restriction to $\mathscr{X}_j$ of the projection $\mathbb{P}(V_{\gamma_j}^*) \times \mathbb{A}^1 \to \mathbb{P}(V_{\gamma_j}^*)$ to the first factor. Then z_j, $j = 1, 2, \cdots, N_j$, are viewed as sections for

$$\mathscr{L}_j = \mathrm{pr}_1^* \mathscr{O}_{\mathbb{P}^{N_j-1}(\mathbb{C})}(1),$$

and $(|z_1|^2 + \cdots + |z_{N_j}|^2)^{-1}$ defines a Hermitian metric $h_{\mathrm{FS}} := (|v_1|^2 + \cdots + |v_{N_j}|^2)^{-1}$ for $\mathscr{L}_j$ such that the associated Ricci form $\mathrm{Ric}(h_{\mathrm{FS}}) := -dd^c \log h_{\mathrm{FS}}$ is the Fubini–Study form ω_{FS} on $\mathscr{X}_j$ over $\mathbb{A}^1$. We then consider the Deligne pairing

$$\phi_j := \langle \phi_{\mathrm{FS}}, \cdots, \phi_{\mathrm{FS}} \rangle_{\mathscr{X}_j/\mathbb{A}^1}$$

over $\mathbb{A}^1$ of $(n+1)$-pieces of Kähler potentials $\phi_{\mathrm{FS}} := -\log h_{\mathrm{FS}}$. Since the blow-up of W_j at the origin is viewed as the tautological line bundle $\mathscr{O}_{\mathbb{P}(W_j)}(-1)$, the Chow norm for W_j induces a Hermitian metric h_{CH} for

$$\bar{\mathscr{N}} := \mathscr{O}_{\mathbb{P}(W_j)}(1).$$

In view of (7.5), we put $Z(t) := \sigma_j(t) \cdot \hat{X}_j$ for $t \in \mathbb{C}^*$. The map sending $t \in \mathbb{C}^*$ to the Chow point $[Z(t)] \in \mathbb{P}(W_j)$ extends to a $\mathbb{C}^*$-equivariant algebraic map

$$\bar{Z} : \mathbb{A}^1 \to \mathbb{P}(W_j).$$

Then by a theorem of Zhang [93], the pullback by $\bar{Z}$ defines an isomorphism

$$\bar{Z}^* : \bar{\mathscr{N}} \cong \mathscr{N}$$

inducing an isometry between $(\bar{\mathscr{N}}, h_{\mathrm{CH}})$ and $(\mathscr{N}, h_j)$, where $h_j := e^{-\phi_j}$. Let $0 \ne w \in \mathscr{N}_1$, where $\mathscr{N}_1$ is the fiber of $\mathscr{N}$ over $1 \in \mathbb{A}^1$. We consider the $\mathbb{C}^*$-action on $\mathscr{N}$ induced by the $\mathbb{C}^*$-action on $\mathscr{L}_j$. Let λ_j be the weight of the $\mathbb{C}^*$-action on the fiber $\mathscr{N}_0$ over the origin. For each $t \in \mathbb{C}^*$, we put

$$v_j(t) := t^{-\lambda_j}\{\sigma_j(t) \cdot w\}. \qquad (7.6)$$

The mapping $\mathbb{A}^1 \setminus \{0\} \ni t \mapsto \hat{v}_j(t) := \sigma_j(t) \cdot w$ defines a nowhere vanishing holomorphic section of $\mathscr{N}$ over $\mathbb{A}^1 \setminus \{0\}$, which extends to a rational section for $\mathscr{N}$ over $\mathbb{A}^1$. Let μ be the order of $\hat{v}$ at the origin, i.e., $\mu = \mathrm{ord}_{z=0} \hat{v}_j(z)$. Then by the same argument as in the proof of Theorem 2.1, we obtain

$$\mu = \lambda_j.$$

Hence the mapping $\mathbb{A}^1 \setminus \{0\} \ni z \mapsto v_j(z) \in \mathcal{N}_z$ extends to a nowhere vanishing section of $\mathcal{N}$ over $\mathbb{A}^1$ (which trivializes $\mathcal{N}$). Since the Chow form $0 \neq \hat{X}_j \in W_j$ is viewed as a point $\neq 0$ in the fiber of $\mathcal{O}_{\mathbb{P}(W_j)}(-1)$ over $[\hat{X}_j] \in \mathbb{P}(W_j)$, we have

$$0 \neq \hat{X}_j^{-1} \in \bar{\mathcal{N}}$$

sitting over $[\hat{X}_j] \in \mathbb{P}(W_j)$. Hence by setting $w := \bar{Z}^* \hat{X}_j^{-1} \neq 0$, we obtain

$$\|\sigma_j(t) \cdot \hat{X}_j\|_{\mathrm{CH}(\rho_{\gamma_j})}^{-1} = \|\sigma_j(t) \cdot \hat{X}_j^{-1}\|_{h_{\mathrm{CH}}} = \|\sigma_j(t) \cdot w\|_{h_j} = t^{\lambda_j} \|v_j(t)\|_{h_j}, \quad t \in \mathbb{R}_+,$$

in terms of the notation (7.6). Put $t = \exp \tilde{s}$ for $\tilde{s} \in \mathbb{R}$. Then from the equality just above, it follows that

$$\log \|\sigma_j(t) \cdot \hat{X}_j\|_{\mathrm{CH}(\rho_{\gamma_j})} = -\lambda_j \tilde{s} - \log \|v_j(t)\|_{h_j}. \tag{7.7}$$

By $v_j(0) \neq 0$ together with a result of Moriwaki [60], we see that $\|v_j(z)\|_{h_j}$ is a positive continuous function in a neighborhood of the origin. Then by the same argument as in (2.11) and (2.12),

$$\lim_{\tilde{s} \to -\infty} \frac{d}{d\tilde{s}} \log \|\sigma_j(t) \cdot \hat{X}_j\|_{\mathrm{CH}(\rho_{\gamma_j})} = -\lambda_j. \tag{7.8}$$

We now put $\tilde{s} := s/|u_j|_\infty$. By convexity of the left-hand side of (7.7) as a function of $\tilde{s}$, we see that $\kappa_j := -\log \|v_j(t)\|_{h_j}$ is also a convex function $\kappa_j(s)$ of s, so that its derivative $\dot{\kappa}_j(s)$ with respect to s is a nonnegative function satisfying

$$\lim_{s \to -\infty} \dot{\kappa}_j(s) = 0.$$

Since $t = \exp(s/|u_j|_\infty)$, in terms of the function $f_j(s)$ at the beginning of this section, we can rewrite (7.7) as follows:

$$f_j(s) = -\frac{1}{|u_j|_1} \gamma_j^{-n} \lambda_j s + \frac{|u_j|_\infty}{|u_j|_1} \gamma_j^{-n} \cdot \kappa_j.$$

By differentiating this equality with respect to s, we obtain the following:

$$\dot{f}_j(s) = -\frac{1}{|u_j|_1} \gamma_j^{-n} \lambda_j + \frac{1}{|u_j|_1} \gamma_j^{-n} \dot{\kappa}_j(s) \geq -\frac{1}{|u_j|_1} \gamma_j^{-n} \lambda_j.$$

Let $j \to \infty$ and then let $s \to -\infty$. Since $\underline{\lim}_{j \to \infty} \dot{f}_j(s)$ is non-decreasing in s, we have

$$F_1(\{\mu_j\}) = \lim_{s \to -\infty} \underline{\lim}_{j \to \infty} \dot{f}_j(s) \geq \phi(\{\mu_j\}), \tag{7.9}$$

where $\phi(\{\mu_j\}) \in \mathbb{R} \cup \{-\infty\}$ is defined by

$$\phi(\{\mu_j\}) := \lim_{j \to \infty} \left(-\frac{1}{|u_j|_1} \gamma_j^{-n} \lambda_j \right).$$

Since $\lim_{s \to -\infty} \dot{f}_j(s) = -|u_j|_1^{-1} \gamma_j^{-n} \lambda_j$ by (7.8), we can write $\phi(\{\mu_j\})$ as

$$\phi(\{\mu_j\}) = \lim_{j \to \infty} \lim_{s \to -\infty} \dot{f}_j(s).$$

Hence $F_1(\{\mu_j\}) = \phi(\{\mu_j\})$ in (7.9) if the double limit in the definition of $F(\{\mu_j\})$ commutes. By comparing (7.8) with (5.14) (or by the definition of λ_j), we see that

$$-\lambda_j = \lambda_{\sigma_j}, \tag{7.10}$$

which is the Chow weight for $X \subset \mathbb{P}(V_{\gamma_j}^*)$ for the $\mathbb{C}^*$-action via σ_j, i.e., the weight of the $\mathbb{C}^*$-action on $\lim_{|t| \to 0} \sigma(t) \hat{X}_j$ via σ_j. The Kodaira embedding and the twisted Kodaira embedding define the same abstract image (see the explanation at the beginning of Sect. 5.1). Hence the weight λ_{σ_j} doesn't change even if we replace the twisted Kodaira embedding by the ordinary Kodaira embedding in the definition of X_j in Sect. 4.5, where $\sigma_j(t), t \in \mathbb{C}^*$, commutes multiplicatively with the diagonal matrix of order N_{γ_j} with the (i, α)-th diagonal element

$$\{1 - \gamma_j^{-2} \chi_{\gamma_j, i}(\sqrt{-1} y_{\gamma_j})\}^{1/2}$$

by the notation in Sect. 4.4. Moreover, the weight λ_{σ_j} has nothing to do with the choice of the reference Hermitian metric h and $\omega := \mathrm{Ric}(h)$. Hence:

Theorem 7.3 *The right-hand side $\phi(\{\mu_j\})$ in (7.9) is independent of the choice of the reference Hermitian metric h for L. Moreover, as far as the value of $\phi(\{\mu_j\})$ is concerned, we may set $X_j = \Phi_{\gamma_j}(X)$ in Sect. 4.5 by replacing the twisted Kodaira embedding by the ordinary Kodaira embedding.*

It is very plausible that the equality $F_1(\{\mu_j\}) = \phi(\{\mu_j\})$ holds. Here by using the inequality (7.9), we give a proof for (5.29) in Problem 5.1.

Proof of (5.29) In Problem 5.1, replacing L by $L^{\otimes \gamma}$, we may assume $\gamma = 1$ without loss of generality. Then by Theorem 2.2, we obtain

$$\lambda_{\sigma_j} = \lim_{s \to -\infty} \dot{f}_j(s) = (n+1)! a_0 \sum_{\alpha=1}^{\infty} \mathrm{DF}_\alpha(\mathscr{X}, \mathscr{L}) j^{n+1-\alpha}, \tag{7.11}$$

where $j \gg 1$. Note that $\gamma_j = j$ by $\gamma = 1$. Since $a_0 = c_1(L)^n[X]/n!$, it follows from (7.9), (7.10) and (7.11) that

$$F_1(\{\mu_j\}) \geq \phi(\{\mu_j\}) = \lim_{j \to \infty} \left(\frac{1}{|u_j|_1} j^{-n}\lambda_{\sigma_j}\right)$$

$$= \lim_{j \to \infty} \frac{(n+1)c_1(L)^n[X] \sum_{\alpha=1}^{\infty} \mathrm{DF}_\alpha(\mu) \, j^{1-\alpha}}{|u_j|_1} = \frac{(n+1)c_1(L)^n[X]}{\|\mu\|_{asymp}} \mathrm{DF}_1(\mu),$$

where the last equality follows from $\lim_{j \to \infty} |u_j|_1 = \|\mu\|_{asymp}$ (cf. Sect. 4.2).

Problems

7.1 For a polarized algebraic manifold (X, L), assume that X admits a CSC Kähler metric in the class $c_1(L)$. Show that (X, L) is K-semistable.

7.2 (cf. [66]) Let (X, L) be a polarized algebraic manifold such that X admits a CSC Kähler metric in the class $c_1(L)$. Is (X, L) always asymptotically Chow stable?

Chapter 8
Existence Problem

Abstract In this chapter, we give some remarks on the existence of extremal Kähler metrics on polarized algebraic manifolds:

- For the existence of extremal Kähler metrics, as mentioned in Sect. 8.1, a result of He states the following: a polarized algebraic manifold (X, L) admits an extremal Kähler metric in the class $c_1(L)$ if the modified K-energy is proper modulo the action of the centralizer G_0 of the extremal vector field in $\mathrm{Aut}^0(X)$. Hence the existence of an extremal Kähler metric is reduced to showing the properness of the modified K-energy modulo the action of G_0.
- In Sect. 8.2, we give some observations on the existence of extremal metrics, which gives an idea of how strong K-stability is useful for the existence.

Keywords The theorem of Chen, Donaldson and Sun and of Tian · The results of Chen, Cheng and He

8.1 A Result of He on the Existence of Extremal Kähler Metrics

A Kähler–Einstein metric always exists on a K-polystable Fano manifold by the affirmative solution of the Yau–Tian–Donaldson conjecture (see Conjecture 6.1 in Sect. 6.2) by Chen et al. [13–15] and Tian [80],

Recently, for a general polarization, a breakthrough by Chen and Cheng [10–12] on the existence of CSC Kähler metrics has been developed by He [29] to the case of the existence problem of extremal Kähler metrics. Namely, given a polarized algebraic manifold (X, L), Theorem 9.7 in Sect. 9.6 below is valid even if we replace $c_1(X)$ by an arbitrary polarization class $c_1(L)$, i.e., *X admits an extremal Kähler metric in the class $c_1(L)$ if the modified K-energy $\hat{\kappa}$ is proper modulo the action of the centralizer G_0 of the extremal vector field in* $\mathrm{Aut}^0(X)$.

T. Mabuchi, *Test Configurations, Stabilities and Canonical Kähler Metrics*,
SpringerBriefs in Mathematics, https://doi.org/10.1007/978-981-16-0500-0_8

8.2 Some Observations on the Existence Problem

We consider here a polarized algebraic manifold (X, L) which is strongly K-stable relative to a maximal algebraic torus $T_{\max}$ in $\mathrm{Aut}^0(X)$. For simplicity, assume that

$$T_{\max} = \{1\}.$$

Fix a Hermitian metric h for L such that $\omega := \mathrm{Ric}(h)$ is Kähler. Then by Theorem 5.5, (X, L) is asymptotically Chow stable, i.e., there exists an increasing sequence

$$1 < \gamma_1 < \gamma_2 < \cdots < \gamma_j < \cdots$$

with a balanced metric $\omega_j = \mathrm{Ric}(h_j)$ for $L^{\otimes \gamma_j}$, $j = 1, 2, \cdots$, where h_j is a Hermitian metric for L. Hence we have an orthonormal basis $\{v_1, v_2, \cdots, v_{N_j}\}$ for $(V_{\gamma_j}, \rho_{\gamma_j})$ such that $(|v_1|^2 + |v_2|^2 + \cdots + |v_{N_j}|^2)_{h_j}$ is a constant, where $\rho_j = \rho_{\gamma_j}$ is such that

$$\rho_{\gamma_j}(v, v') = \int_X (v, v')_{h_j} \omega_j^n, \qquad v, v' \in V_{\gamma_j}.$$

On the other hand, another Hermitian inner product ρ_0 for V_{γ_j} is defined by

$$\rho_0(v, v') = \int_X (v, v')_h \omega^n, \qquad v, v' \in V_{\gamma_j}.$$

By choosing a suitable orthonormal basis $\{w_1, w_2, \cdots, w_{N_j}\}$ for (V_{γ_j}, ρ_0), and by replacing h_j by constant times h_j if necessary, we can view ρ_0 as an $N_j \times N_j$ identity matrix, and ρ_j is also written as an $N_j \times N_j$ diagonal matrix in $\mathrm{SL}(N_j; \mathbb{C})$ with each α-th diagonal element $\lambda_\alpha > 0$. Put $\underline{b}_\alpha := (1/2) \log \lambda_\alpha$. Then $\underline{b}_1 + \underline{b}_2 + \cdots + \underline{b}_{N_j} = 0$. Approximating each $\underline{b}_\alpha$ by a sequence of rational numbers, we may assume from the beginning that each $\underline{b}_\alpha$ is a rational number (see [53]). By choosing a positive integer m_j, we have that

$$b_\alpha := m_j \underline{b}_\alpha, \qquad \alpha = 1, 2, \cdots, N_j,$$

are all integers. Let u_j (reps. $\underline{u}_j$) in $\mathfrak{sl}(V_{\gamma_j})$ be the diagonal matrices with each α-th diagonal element b_α (resp. $\underline{b}_\alpha$). Let $\sigma_j : \mathbb{C}^* \to \mathrm{SL}(V_{\gamma_j})$ be the special one-parameter group with the fundamental generator u_j, so that $\sigma_j(t) = t^{u_j}$. In the definition

$$f_j(s) := \frac{|u_j|_\infty}{|u_j|_1} \gamma_j^{-n} \log \| \exp(s u_j / |u_j|_\infty) \cdot \hat{X}_j \|_{\mathrm{CH}(\rho_{\gamma_j})}$$

of $f_j(s)$, the function $f_j(s)$ doesn't change even if we replace u_j by $\underline{u}_j$, and hence the function $f_j(s)$ is rewritten as

$$f_j(s) := \frac{|\underline{u}_j|_\infty}{|\underline{u}_j|_1}\gamma_j^{-n}\log\|\exp(s\underline{u}_j/|\underline{u}_j|_\infty)\cdot\hat{X}_j\|_{\mathrm{CH}(\rho_{\gamma_j})}.$$

On the other hand, $|\underline{u}_j|_\infty$ (see (4.1)) can be viewed as the distance between ρ_{γ_j} and ρ_0. Let $\mu_j := (\mathscr{X}_{\sigma_j},\mathscr{L}_{\sigma_j})$ be the test configuration associated to σ_j. Put

$$d_\infty := \sup_j |\underline{u}_j|_\infty.$$

Then the following cases are possible:

(Case 1) $d_\infty = +\infty$;
(Case 2) $d_\infty < +\infty$.

We first consider Case 1. Note that $\sigma_j(t) = \exp(s\underline{u}_j/|\underline{u}_j|_\infty)$, $t = \exp(s/|u_j|_\infty)$. Here the α-th diagonal element $\underline{b}_\alpha$ of the diagonal matrix $\underline{u}_j$ is $(1/2)\log\lambda_\alpha$. Hence

$$\sigma_j(t)\cdot\rho_0{}_{|s=0} = \rho_0,$$

$$\sigma_j(t)\cdot\rho_0{}_{|s=-|\underline{u}_j|_\infty} = \exp(-\underline{u}_j)\cdot\rho_0,$$

where $\exp(-\underline{u}_j)$ is a diagonal matrix with α-th diagonal element $e^{-\underline{b}_\alpha}$, and it acts on $V_{\gamma_j}^*$ by the contragradient representation. Since ρ_0 sitting in $V_{\gamma_j}^* \otimes \bar{V}_{\gamma_j}^*$ is written as an $N_j \times N_j$ identity matrix I in terms of the basis $\{w_1, w_2, \cdots, w_{N_j}\}$, we can view $\sigma_j(t)\cdot\rho_0{}_{|s=-|\underline{u}_j|_\infty}$ as a diagonal matrix

$$^t\exp(\underline{u}_j)\cdot I\cdot\exp(\underline{u}_j)$$

whose α-th diagonal element is $e^{2\underline{b}_\alpha} = \lambda_\alpha$. Thus we obtain

$$\sigma_j(t)\cdot\rho_0{}_{|s=-|\underline{u}_j|_\infty} = \rho_j.$$

Since the Chow norm takes the critical value at the balanced metric ω_j (which corresponds to ρ_j), the derivative of the function $f_j(s)$ at $s = -|\underline{u}_j|_\infty$ vanishes:

$$\dot{f}_j(-|\underline{u}_j|_\infty) = 0 \qquad \text{for all } j.$$

By the assumption of Case 1, $d_\infty = +\infty$. Hence replacing $\{\underline{u}_j\}$ by its subsequence if necessary, we may assume that

$$|\underline{u}_j|_\infty, \qquad j = 1, 2, \cdots,$$

is a monotone increasing sequence diverging to $+\infty$. For the time being, fix an arbitrary j. Then by the monotonicity of the sequence,

$$|\underline{u}_{j'}|_\infty \geq |\underline{u}_j|_\infty \quad \text{for all } j' \geq j.$$

Hence $-|\underline{u}_{j'}|_\infty \leq -|\underline{u}_j|_\infty$. Since $\dot{f}_{j'}(s)$ is a non-decreasing function of s, we obtain

$$0 = \dot{f}_{j'}(-|\underline{u}_{j'}|_\infty) \leq \dot{f}_{j'}(-|\underline{u}_j|_\infty) \quad \text{for all } j' \geq j.$$

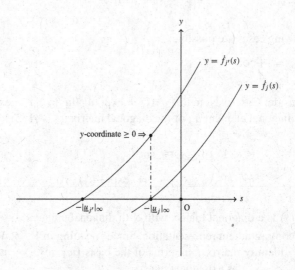

By letting $j' \to \infty$, we obtain $\varliminf_{j'\to\infty} \dot{f}_{j'}(s) \geq 0$ for all s such that $-|\underline{u}_j|_\infty \leq s < +\infty$. Now let $j \to \infty$. Then by $|\underline{u}_j|_\infty \to +\infty$, it follows that

$$\varliminf_{j'\to\infty} \dot{f}_{j'}(s) \geq 0 \quad \text{for all } s \text{ with } -\infty < s < +\infty.$$

Here we let $s \to -\infty$. Then

$$F_1(\{\mu_j\}) = \lim_{s\to-\infty} \varliminf_{j'\to\infty} \dot{f}_{j'}(s) \geq 0.$$

However, by $|\underline{u}_j|_\infty \to +\infty$, we may assume that all μ_j are nontrivial, and therefore it follows from the strong K-stability of (X, L) that the inequality $F_1(\{\mu_j\}) < 0$ holds in contradiction. Hence Case 1 cannot occur.

We now consider Case 2. In this case by $d_\infty < +\infty$, we intuitively see that ρ_j is not so far away from ρ_0 (and hence ω_j is not so far away from ω). For instance we have an a priori C^0 bound for the sequence of balanced metrics (cf. [53]). It is then expected that ω_j converges to a CSC Kähler metric.

The above argument goes through also for relative versions including the case of the existence problem of extremal Kähler metrics.

Problems

8.1 Recall the following theorem of Matsushima [59]:

Theorem *Let G be a connected linear algebraic group and let H be its closed algebraic subgroup. If G is reductive, then G/H is affine if and only if H is reductive.*

In view of this theorem, what can you say about the relationship between the asymptotic Chow polystability for (X, L) and the Matsushima–Lichnerowicz obstruction (cf. [38, 58]) for the existence of CSC Kähler metrics in the class $c_1(L)$?

8.2 Since $\mathbb{P}^2(\mathbb{C})$ and $\mathbb{P}^1(\mathbb{C}) \times \mathbb{P}^1(\mathbb{C})$ are homogeneous spaces, they admit Kähler–Einstein metrics (the Fubini–Study metric or a product of the Fubini–Study metrics). Enumerate all other compact complex connected surfaces with Kähler–Einstein metrics of positive Ricci curvature.

Chapter 9
Canonical Kähler Metrics on Fano Manifolds

Abstract In this chapter, we study canonical Kähler metrics in the class $c_1(X)$ for Fano manifolds with no Kähler–Einstein metrics. Typical examples are:

- Kähler–Ricci solitons,
- Extremal Kähler metrics,
- Generalized Kähler–Einstein metrics.

Since Kähler-Ricci solitons and extremal Kähler metrics are well-known, we focus on the recent developments of the studies of generalized Kähler–Einstein metrics.

- In Sects. 9.1 and 9.2, we discuss basic facts concerning holomorphic vector fields (including the extremal vector field) on Fano manifolds.
- In Sects. 9.3 and 9.4, we discuss obstructions to the existence of generalized Kähler–Einstein metrics.
- In Sects. 9.5, 9.7, 9.8, and 9.9, we discuss the existence problem of generalized Kähler–Einstein metrics including both Yao's result and Hisamoto's result.
- In Sect. 9.6, we shall show that a generalized Kähler–Einstein manifold always admits an extremal Kähler metric in the class $c_1(X)$.

Keywords Generalized Kähler–Einstein metrics · Extremal vector fields · Yao's result · Hisamoto's result

9.1 Kähler Metrics in Anticanonical Class

A compact complex connected manifold X is called a *Fano* manifold if $c_1(X) > 0$. For an n-dimensional Fano manifold X, let ω be a Kähler form in the class $c_1(X)$. Then the corresponding *Ricci potential* is the real-valued smooth function $f_\omega \in C^\infty(X)_\mathbb{R}$ on X defined uniquely by

$$dd^\mathbb{C} f_\omega = \text{Ric}(\omega) - \omega \quad \text{and} \quad \int_X e^{f_\omega} \omega^n = \int_X \omega^n.$$

© The Author(s), under exclusive licence to Springer Nature Singapore Pte Ltd. 2021   75
T. Mabuchi, *Test Configurations, Stabilities and Canonical Kähler Metrics*,
SpringerBriefs in Mathematics, https://doi.org/10.1007/978-981-16-0500-0_9

Since $c_1(X) > 0$, by the Kodaira vanishing theorem, we obtain

$$h^{1,0}(X) = h^{0,1}(X) = h^{n-1}(X, \mathscr{O}(K_X)) = 0.$$

Hence all holomorphic vector fields on the Kähler manifold (X, ω) are Hamiltonian (cf. Sect. 3.2) in the sense that the Lie algebra

$$\mathfrak{g}_\omega := \left\{ \psi \in C^\infty(X)_\mathbb{C}; \ \operatorname{grad}_\omega^\mathbb{C} \psi \text{ is holomorphic and } \int_X \psi \omega^n = 0 \right\}$$

endowed with the Poisson bracket is identified with the space $\mathfrak{g} := H^0(X, \mathscr{O}(TX))$ of holomorphic vector fields on X by the isomorphism

$$\mathfrak{g}_\omega \cong \mathfrak{g}, \qquad \psi \leftrightarrow \operatorname{grad}_\omega^\mathbb{C} \psi.$$

Definition 9.1 (cf. [45]) ω above is called *generalized Kähler–Einsten* if the complex vector field $\operatorname{grad}_\omega^\mathbb{C}(1 - e^{f_\omega})$ is holomorphic, i.e., $1 - e^{f_\omega} \in \mathfrak{g}_\omega$.

Recall that ω is *Kähler–Einstein* if f_ω is constant, whereas ω is called a *Kähler–Ricci soliton* if $\operatorname{grad}_\omega^\mathbb{C} f_\omega$ is holomorphic. Obviously, a Kähler–Einstein form ω is always generalized Kähler–Einstein. Recall also that ω is called *extremal Kähler* if the complex vector field $\operatorname{grad}_\omega^\mathbb{C} S_\omega (= \operatorname{grad}_\omega^\mathbb{C}(S_\omega - n))$ associated to the scalar curvature function S_ω of ω is holomorphic.

9.2 Extremal Vector Fields

In this section, we start from a general situation that (X, L) is a polarized algebraic manifold such that L is an ample line bundle which is not necessarily very ample. Here X is not necessarily a Fano manifold, nor is L necessarily the anticanonical bundle K_X^{-1}. For a Kähler metric ω in the class $c_1(L)$, we put

$$\mathfrak{k}_\omega := \{ \psi \in \mathfrak{g}_\omega; \ \psi \text{ is real-valued} \},$$

where $\mathfrak{g}_\omega$ is as defined in Sect. 9.1. Then for the space $\mathfrak{k} := \{ \operatorname{grad}_\omega^\mathbb{C} \psi; \ \psi \in \mathfrak{k}_\omega \}$ of the Killing vector fields on (X, ω), we have the isomorphism $\mathfrak{k}_\omega \cong \mathfrak{k}$. Let $\mathfrak{k}_\omega^\perp$ be the orthogonal complement

$$\left\{ \varphi \in L^2(X, \omega)_\mathbb{R}; \ \int_X \varphi \psi \, \omega^n = 0 \text{ for all } \psi \in \mathfrak{k}_\omega \right\}$$

of $\mathfrak{k}_\omega$ in the space $L^2(X, \omega)_\mathbb{R}$ of all real-valued L^2 functions on the Kähler manifold (X, ω). For the projection $\operatorname{pr} : L^2(X, \omega)_\mathbb{R} (= \mathfrak{k}_\omega \oplus \mathfrak{k}_\omega^\perp) \to \mathfrak{k}_\omega$ to the first factor, by viewing $\operatorname{pr}(S_\omega - n)$ as an element y_ω of $\mathfrak{k}$ by the isomorphism $\mathfrak{k} \cong \mathfrak{k}_\omega$, the holomorphic vector field $y_\omega \in \mathfrak{k}$ on X is called the *extremal vector field* on (X, L) for ω. Then by [26], the associated real vector field $y_\omega^{\mathbb{R}} := y_\omega + \bar{y}_\omega$ on X satisfies

$$\exp(2\pi m y_\omega^{\mathbb{R}}) = \mathrm{id}_X \tag{9.1}$$

for some positive integer m, where $y_\omega = 0$ if and only if the Futaki character $\mathscr{F}$ (cf. Sect. 3.1) vanishes. If $y_\omega \neq 0$, then $y_\omega^{\mathbb{R}}$ generates an isometric S^1-action on (X, ω). From now on, we assume that $L = K_X^{-1}$, i.e., the Kähler class $c_1(L)$ is $c_1(X)$.

Theorem 9.1 $\mathrm{pr}(S_\omega - n) = \mathrm{pr}(1 - e^{f_\omega})$.

Proof For the Laplacian $\Delta_\omega := \sum_{\alpha,\beta} g^{\bar{\beta}\alpha} \partial^2/\partial z_\alpha \partial z_{\bar{\beta}}$, we put

$$D_\omega := \Delta_\omega + \sum_{\alpha,\beta} g^{\bar{\beta}\alpha} \frac{\partial f_\omega}{\partial z_\alpha} \frac{\partial}{\partial z_{\bar{\beta}}}.$$

Then a formula in [24, p. 41], shows that

$$\int_X (D_\omega \varphi_1)\bar{\varphi}_2\, e^{f_\omega}\omega^n = \int_X \varphi_1\overline{(D_\omega\varphi_2)}\, e^{f_\omega}\omega^n = -\int_X (\bar{\partial}\varphi_1, \bar{\partial}\varphi_2)_\omega\, e^{f_\omega}\omega^n$$

for all $\varphi_1, \varphi_2 \in C^\infty(X)_{\mathbb{C}}$, where $(\ ,\ )_\omega$ denotes the pointwise Hermitian inner product by ω for 1-forms on X. Moreover,

$$\mathrm{Ker}_{\mathbb{R}}(D_\omega + 1) \cong \mathfrak{k}_\omega, \qquad \varphi \leftrightarrow \varphi - C_\varphi, \tag{9.2}$$

where $\mathrm{Ker}_{\mathbb{R}}(D_\omega + 1) := \{\varphi \in C^\infty(X)_{\mathbb{R}};\ (D_\omega + 1)\varphi = 0\}$, and $C_\varphi := \int_X \varphi\omega^n/\int_X \omega^n$. Hence for every $\psi \in \mathfrak{k}_\omega$, there exists $\varphi \in \mathrm{Ker}_{\mathbb{R}}(D_\omega + 1)$ such that $\psi = \varphi - C_\varphi$. Then

$$\int_X \psi\, \mathrm{pr}(1 - e^{f_\omega})\omega^n = \int_X \psi(1 - e^{f_\omega})\omega^n = \int_X (\varphi - C_\varphi)(1 - e^{f_\omega})\omega^n$$

$$= \int_X \varphi(1 - e^{f_\omega})\omega^n = \int_X (-D_\omega\varphi)(1 - e^{f_\omega})\omega^n = -\int_X (D_\omega\varphi)\omega^n + \int_X (D_\omega\varphi)e^{f_\omega}\omega^n$$

$$= -\int_X (D_\omega\varphi)\omega^n + \int_X \varphi\overline{(D_\omega 1)}\, e^{f_\omega}\omega^n = -\int_X (D_\omega\varphi)\omega^n$$

$$= -\int_X \left(\Delta_\omega\varphi + \sum_{\alpha,\beta} g^{\bar{\beta}\alpha} \frac{\partial f_\omega}{\partial z_\alpha} \frac{\partial\varphi}{\partial z_{\bar{\beta}}}\right)\omega^n = -\int_X \left(\sum_{\alpha,\beta} g^{\bar{\beta}\alpha} \frac{\partial f_\omega}{\partial z_\alpha} \frac{\partial\varphi}{\partial z_{\bar{\beta}}}\right)\omega^n$$

$$= -\int_X (\bar{\partial}\varphi, \bar{\partial} f_\omega)_\omega\, \omega^n = -\int_X (\bar{\partial}\psi, \bar{\partial} f_\omega)_\omega\, \omega^n = \int_X \psi(\Delta_\omega f_\omega)\, \omega^n$$

$$= \int_X \psi(S_\omega - n)\, \omega^n = \int_X \psi\, \mathrm{pr}(S_\omega - n)\, \omega^n.$$

Since both $\text{pr}(1 - e^{f_\omega})$ and $\text{pr}(S_\omega - n)$ are in $\mathfrak{k}_\omega$, it therefore follows that $\text{pr}(1 - e^{f_\omega})$ coincides with $\text{pr}(S_\omega - n)$, as required.   □

9.3 An Obstruction of Matsushima's Type

If X admits a generalized Kähler–Einstein metric, we have a decomposition theorem of the Lie algebra $\mathfrak{g}$ as shown for extremal Kähler metrics by Calabi [9]. For each nonnegative rational number λ, we consider the linear subspace $\mathfrak{g}_\lambda$ of $\mathfrak{g}$ defined by

$$\mathfrak{g}_\lambda := \{ u \in \mathfrak{g} ; [\sqrt{-1}\, y_\omega, u] = \lambda u \}.$$

Then for $\lambda = 0$, the associated Lie subalgebra $\mathfrak{g}_0$ of $\mathfrak{g}$ is nothing but the centralizer $Z_\mathfrak{g}(y_\omega)$ of the extremal vector field y_ω in $\mathfrak{g}$. Let $\mathfrak{k}^\mathbb{C}$ be the complexification of $\mathfrak{k}$ in $\mathfrak{g}$.

Theorem 9.2 *If X admits a generalized Kähler–Einstein metric, then $\mathfrak{g}_0$ coincides with $\mathfrak{k}^\mathbb{C}$, and there exists a sequence of rational numbers $0 = \lambda_0 < \lambda_1 < \cdots < \lambda_r$ such that $\mathfrak{g}$ viewed as a vector space is written as a direct sum*

$$\mathfrak{g} = \bigoplus_{i=0}^{r} \mathfrak{g}_{\lambda_i}. \tag{9.3}$$

Proof By $h^{1,0}(X) = 0$, the identity component $G := \text{Aut}^0(X)$ of the group of all holomorphic automorphisms of X is linear algebraic. Then the Chevalley decomposition allows us to write the associated Lie algebra $\mathfrak{g}$ as a semi-direct sum

$$\mathfrak{g} = \mathfrak{k}^\mathbb{C} \ltimes \mathfrak{u},$$

for the unipotent radical $\mathfrak{u}$ of $\mathfrak{g}$. Let T be the algebraic torus generated by y_ω. Then by setting $t = e^s$ for $s \in \mathbb{C}$, we have an isomorphism $i : \mathbb{C}^* \cong T$ such that

$$i(t) := \exp(s\sqrt{-1}\, m y_\omega),$$

for m as in (9.1). Since the T-action on $\mathfrak{u}$ by the adjoint representation is completely reducible, an increasing sequence of integers $k_1 < k_2 < \cdots < k_r$ exists such that $\mathfrak{u}$ viewed as a vector space is a direct sum $\mathfrak{u}_1 \oplus \mathfrak{u}_2 \oplus \cdots \oplus \mathfrak{u}_r$, where

$$\mathfrak{u}_\alpha := \{ u \in \mathfrak{u}; \text{Ad}(i(t))(u) = t^{k_\alpha} u \} = \{ u \in \mathfrak{u}; [\sqrt{-1}\, m y_\omega, u] = k_\alpha u \},$$

for $\alpha = 1, 2, \ldots, r$. Then by setting $\lambda_\alpha := k_\alpha / m$, we can write

$$\mathfrak{u}_\alpha = \{ u \in \mathfrak{u}; [\sqrt{-1}\, y_\omega, u] = \lambda_\alpha u \} = \{ u \in \mathfrak{g}; [\sqrt{-1}\, y_\omega, u] = \lambda_\alpha u \}. \tag{9.4}$$

Recall that y_ω sits in the center of $\mathfrak{k}^{\mathbb{C}}$ (cf. [26]). Hence $\mathfrak{k}^{\mathbb{C}} \subset Z_{\mathfrak{g}}(y_\omega)$. Then the proof is reduced to showing that

$$\lambda_\alpha > 0, \qquad \alpha = 1, 2, \ldots, r. \tag{9.5}$$

Because if (9.5) is true, then by $\mathfrak{g} = \mathfrak{k}^{\mathbb{C}} \ltimes \mathfrak{u}$, the equality (9.3) is straightforward from the decomposition $\mathfrak{u} = \mathfrak{u}_1 \oplus \mathfrak{u}_2 \oplus \cdots \oplus \mathfrak{u}_r$. Furthermore, by writing each element w in $Z_{\mathfrak{g}}(y_\omega)$ as a sum $\kappa + u$ for some $\kappa \in \mathfrak{k}^{\mathbb{C}}$ and $u \in \mathfrak{u}$, we immediately see from (9.5) and $[\sqrt{-1}y_\omega, w] = 0$ that the equality $u = 0$ holds, i.e., $Z_{\mathfrak{g}}(y_\omega) \subset \mathfrak{k}^{\mathbb{C}}$. Thus we obtain the equality $Z_{\mathfrak{g}}(y_\omega) = \mathfrak{k}^{\mathbb{C}}$, as required.

We shall now show (9.5). Let $0 \neq u \in \mathfrak{u}_\alpha$. For C_φ as in (9.2), we have the following isomorphisms (cf. [24]):

$$\mathrm{Ker}_{\mathbb{C}}(D_\omega + 1) \cong \mathfrak{g}_\omega \cong \mathfrak{g}, \qquad \varphi \leftrightarrow \varphi - C_\varphi \leftrightarrow \mathrm{grad}_\omega^{\mathbb{C}} \varphi,$$

where $\mathrm{Ker}_{\mathbb{C}}(D_\omega + 1)$ is the space of all $\varphi \in C^\infty(X)_{\mathbb{C}}$ such that $(D_\omega + 1)\varphi = 0$. Hence there exist $\varphi_1, \varphi_2 \in \mathrm{Ker}_{\mathbb{C}}(D_\omega + 1)$ such that $u = \mathrm{grad}_\omega^{\mathbb{C}} \varphi_1$ and $y_\omega = \mathrm{grad}_\omega^{\mathbb{C}} \varphi_2$. Since ω is generalized Kähler–Einstein, we obtain

$$1 - e^{f_\omega} = \mathrm{pr}(1 - e^{f_\omega}) = \varphi_2 - C_{\varphi_2}. \tag{9.6}$$

First by $u \neq 0$, we obtain $\bar{\varphi}_1 \notin \mathrm{Ker}_{\mathbb{C}}(D_\omega + 1)$; because if $\bar{\varphi}_1 \in \mathrm{Ker}_{\mathbb{C}}(D_\omega + 1)$, then by $\varphi_1 \in \mathrm{Ker}_{\mathbb{C}}(D_\omega + 1)$, both the real part $\mathrm{Re}\,\varphi_1$ and the imaginary part $\mathrm{Im}\,\varphi_1$ of φ_1 would belong to $\mathrm{Ker}_{\mathbb{R}}(D_\omega + 1)$, implying that $u \in \mathfrak{k}^{\mathbb{C}}$, in contradiction. Secondly,

$$\int_X \bar{\varphi}_1 \, e^{f_\omega} \omega^n = \overline{\int_X \varphi_1 \, e^{f_\omega} \omega^n} = -\overline{\int_X (D_\omega \varphi_1) \, e^{f_\omega} \omega^n} = 0 \tag{9.7}$$

because $\int_X (D_\omega \varphi_1) \, e^{f_\omega} \omega^n = \int_X \varphi_1 \overline{D_\omega(1)} \, e^{f_\omega} \omega^n = 0$. Recall that, by Futaki [24], all eigenvalues of the operator $-D_\omega$ are nonnegative real numbers, where its first positive eigenvalue is 1. Moreover, all eigenfunctions of $-D_\omega$ with eigenvalue 0 are constant. Hence by $\bar{\varphi}_1 \notin \mathrm{Ker}_{\mathbb{C}}(D_\omega + 1)$ and (9.7),

$$\int_X (-D_\omega \bar{\varphi}_1)\varphi_1 \, e^{f_\omega} \omega^n > \int_X |\varphi_1|^2 e^{f_\omega} \omega^n. \tag{9.8}$$

On the other hand, by $\varphi_1 \in \mathrm{Ker}_{\mathbb{C}}(D_\omega + 1)$,

$$\int_X (-D_\omega \varphi_1)\bar{\varphi}_1 \, e^{f_\omega} \omega^n = \int_X |\varphi_1|^2 e^{f_\omega} \omega^n. \tag{9.9}$$

Since the left-hand side $\int_X (-D_\omega \bar{\varphi}_1) \varphi_1 \, e^{f_\omega} \omega^n$ of (9.8) is $\int_X \bar{\varphi}_1 (-\bar{D}_\omega \varphi_1) \, e^{f_\omega} \omega^n$, by subtracting (9.9) from (9.8), we see that

$$\int_X \{(D_\omega - \bar{D}_\omega)\varphi_1\}\bar{\varphi}_1 \, e^{f_\omega} \omega^n \; > \; 0. \tag{9.10}$$

Let $[\; , \;]_\omega$ be the Poisson bracket with respect to the Kähler form ω. In view of the definition of the operator D_ω, we obtain

$$\{(D_\omega - \bar{D}_\omega)\varphi_1\}e^{f_\omega} \; = \; e^{f_\omega}\left(\sum_{\alpha,\beta} g^{\bar{\beta}\alpha}\frac{\partial f_\omega}{\partial z_\alpha}\frac{\partial \varphi_1}{\partial z_{\bar{\beta}}} - \sum_{\alpha,\beta} g^{\bar{\alpha}\beta}\frac{\partial f_\omega}{\partial z_{\bar{\alpha}}}\frac{\partial \varphi_1}{\partial z_\beta}\right)$$

$$= \; \sum_{\alpha,\beta} g^{\bar{\beta}\alpha}\frac{\partial e^{f_\omega}}{\partial z_\alpha}\frac{\partial \varphi_1}{\partial z_{\bar{\beta}}} - \sum_{\alpha,\beta} g^{\bar{\alpha}\beta}\frac{\partial e^{f_\omega}}{\partial z_{\bar{\alpha}}}\frac{\partial \varphi_1}{\partial z_\beta} \; = \; \sum_{\alpha,\beta} g^{\bar{\beta}\alpha}\left(\frac{\partial e^{f_\omega}}{\partial z_\alpha}\frac{\partial \varphi_1}{\partial z_{\bar{\beta}}} - \frac{\partial \varphi_1}{\partial z_\alpha}\frac{\partial e^{f_\omega}}{\partial z_{\bar{\beta}}}\right)$$

$$= \; -\sqrt{-1}\,[e^{f_\omega}, \varphi_1]_\omega \; = \; \sqrt{-1}\,[\varphi_2 - C_{\varphi_2}, \varphi_1 - C_{\varphi_1}]_\omega,$$

where in the last equality, we use (9.6). By the isomorphism $\mathfrak{g}_\omega \cong \mathfrak{g}$, it follows from (9.4) that $[\sqrt{-1}(\varphi_2 - C_{\varphi_2}), \varphi_1 - C_{\varphi_1}]_\omega = \lambda_\alpha(\varphi_1 - C_{\varphi_1})$. Hence $\{(D_\omega - \bar{D}_\omega)\varphi_1\}e^{f_\omega} = \lambda_\alpha(\varphi_1 - C_{\varphi_1})$. From this together with (9.10), it follows that

$$\lambda_\alpha \int_X (\varphi_1 - C_{\varphi_1})\bar{\varphi}_1 \omega^n \; > \; 0. \tag{9.11}$$

Now by $C_{\varphi_1} = \int_X \varphi_1 \omega^n / \int_X \omega^n$ and the Schwarz inequality, we obtain

$$\int_X (\varphi_1 - C_{\varphi_1})\bar{\varphi}_1 \omega^n \; = \; \int_X |\varphi_1|^2 \omega^n - C_{\varphi_1}\int_X \bar{\varphi}_1 \omega^n$$

$$= \; \frac{\int_X |\varphi_1|^2 \omega^n \int_X \omega^n - |\int_X \varphi_1 \omega^n|^2}{\int_X \omega^n} \; \geq \; 0.$$

Here by $u \neq 0$, φ_1 is non-constant, so that in the last inequality, the equality cannot occur. Hence by (9.11), we obtain $\lambda_\alpha > 0$, as required.   □

9.4 An Invariant as an Obstruction to the Existence

For a Kähler metric ω in the class $c_1(M)$, let $\mu_\omega : X \to \mathbb{R}$ be the function on X defined by $\mu_\omega(x) := \mathrm{pr}(S_\omega - n)(x) = \mathrm{pr}(1 - e^{f_\omega})(x)$, $x \in X$. Then by $y_\omega = \mathrm{grad}_\omega^{\mathbb{C}} \mu_\omega$,

$$2\pi \, i_{y_\omega}\omega \; = \; \bar{\partial}\mu_\omega,$$

so that μ_ω defines a moment map for the $\mathbb{C}^*$-action on X generated by y_ω, where the vertices of the image of the moment map are $\mathbb{Q}$-rational (cf. [26]). Hence

$$\gamma_X := \max_X \mu_\omega(x) \in \mathbb{Q}$$

is a number independent of the choice of ω in the class $c_1(X)$, where by $\mu_\omega \in \mathfrak{k}_\omega$, the equality $\int_X \mu_\omega \omega^n = 0$ always holds, and hence $\gamma_X \geq 0$. This γ_X then gives an obstruction [45] to the existence of generalized Kähler–Einstein metrics:

Theorem 9.3 *If X admits a generalized Kähler–Einstein metric, then $\gamma_X < 1$.*

Proof Assume that X admits a generalized Kähler–Einstein metric ω. Then $\mu_\omega = \mathrm{pr}(1 - e^{f_\omega}) = 1 - e^{f_\omega}$. Let $x_0 \in X$ be the point at which the function μ_ω attains its maximum. Then $\gamma_X = \mu_\omega(x_0) = 1 - e^{f_\omega(x_0)} < 1$, as required. $\qquad\square$

9.5 Examples of Generalized Kähler–Einstein Metrics

Let N be a k-dimensional Fano manifold with a Kähler–Einstein form ϕ in the class $c_1(N)$ such that $\mathrm{Ric}(\phi) = \phi$. Let L be a holomorphic line bundle over N with a Hermitian metric h for L such that all eigenvalues $\beta_1, \beta_2, \ldots, \beta_k$ of $\mathrm{Ric}(h)$ with respect to ϕ are constant satisfying $-1 < \beta_i < 1$ for all $i = 1, 2, \cdots, k$. For the vector bundle $\mathscr{O}_N \oplus L$ of rank 2 over N, we consider the associated $\mathbb{P}^1$-bundle

$$X := \mathbb{P}(\mathscr{O}_N \oplus L) \qquad (9.12)$$

over N consisting of all lines through the origin in the fibers of the vector bundle. Hence $n := \dim X$ is nothing but $k + 1$. For the line subbundles $\mathscr{O}_N \oplus \{0\}$, $\{0\} \oplus L$ of E, we put $D_0 := \mathbb{P}(\mathscr{O}_N \oplus \{0\})$, $D_\infty := \mathbb{P}(\{0\} \oplus L)$, respectively. Then $D_0 \cong N$ and $D_\infty \cong N$. Moreover, we have the identification

$$L = X \setminus D_\infty. \qquad (9.13)$$

Let $\rho : L \ni \ell \mapsto \rho(\ell) := \|\ell\|_h \in \mathbb{R}_{\geq 0}$ be the Hermitian norm for L induced by h. Then by (9.13), ρ is viewed as a function on $X \setminus D_\infty$. Define $x \in C^\infty(X \setminus D_0 \cup D_\infty)_\mathbb{R}$ by

$$x = \log(\rho^2).$$

Let $p : X \to N$ be the natural projection. For some real-valued smooth function $\lambda(x)$ of x, we consider the volume form on $X \setminus D_0 \cup D_\infty$,

$$\Omega := \frac{\sqrt{-1}}{2\pi} n \, e^{-\lambda(x)} (p^*\phi)^k \wedge \partial x \wedge \bar\partial x, \qquad (9.14)$$

which is supposed to extend to a volume form on X such that $\int_X \Omega = c_1(X)^n[X]$. Put $\omega := \mathrm{Ric}\,\Omega$. Then by (9.14) and $\mathrm{Ric}(\phi) = \phi$, the same computation as in [43] allows us to write ω as $p^*\phi - \lambda'(x)\,\mathrm{Ric}(h) + \{\sqrt{-1}/(2\pi)\}\,\lambda''(x)\partial x \wedge \bar{\partial}x$. In particular,

$$\omega^n = \frac{\sqrt{-1}}{2\pi}\,n\left\{\lambda''(x)\prod_{i=1}^{k}(1 - \beta_i\lambda'(x))\right\}(p^*\phi)^k \wedge \partial x \wedge \bar{\partial}x. \tag{9.15}$$

Since $dd^c f_\omega = \mathrm{Ric}(\omega) - \omega = dd^c \log(\Omega/\omega^n)$, and since $\int_X e^{f_\omega}\omega^n = \int_X \omega^n = c_1(X)^n[X] = \int_X \Omega = \int_X (\Omega/\omega^n)\omega^n$, we have $e^{f_\omega} = \Omega/\omega^n$. By (9.14) and (9.15),

$$e^{f_\omega} = \frac{\Omega}{\omega^n} = e^{-\lambda(x)}\left\{\lambda''(x)\prod_{i=1}^{k}(1 - \beta_i\lambda'(x))\right\}^{-1}. \tag{9.16}$$

Note that the $\mathbb{C}^*$-action on the line bundle L extends naturally to a holomorphic $\mathbb{C}^*$-action on $X = \mathbb{P}(\mathscr{O}_N \oplus L)$ induced by

$$\mathbb{C}^* \times (\mathscr{O}_N \oplus L) \to \mathscr{O}_N \oplus L, \qquad (t, x \oplus y) \mapsto x \oplus ty.$$

Then $\lambda'(x)$, mapping X onto the interval $[-1, 1]$, defines a moment map for the $\mathbb{C}^*$-action on X. Actually, we have $\lambda'(x) \in \mathrm{Ker}_{\mathbb{R}}(D_\omega + 1)$ and $u := \mathrm{grad}_\omega^{\mathbb{C}}\lambda'(x) \in \mathfrak{k}$ (see for instance [43]). Hence, in order to show that ω is a generalized Kähler–Einstein metric on X, it suffices to solve the following differential equation in $\lambda = \lambda(x)$:

$$\text{R.H.S. of (9.16)} = C_1 + C_2\lambda'(x), \tag{9.17}$$

for some real constants C_1 and C_2 such that Ω extends to a volume form on X and that ω extends to a Kähler form on X. Put

$$b_\alpha := \int_{-1}^{1} q^\alpha \prod_{i=1}^{k}(1 - \beta_i\,q)\,dq, \qquad \alpha = 0, 1, 2.$$

Since $1 - \beta_i\,q$ is positive for all i with $-1 < q < 1$, both b_0 and b_2 are positive. Then by the Schwarz inequality, $b_1^2 < b_0 b_2$. Let z be a system of holomorphic local coordinates on N centered at a point p in N. For a local base $\sigma = \sigma(z)$ for L on a neighborhood of p, by writing $\ell = t\sigma(z)$ for fiber coordinate t for L, we have

$$e^x = \rho^2 = \|\ell\|_h^2 = a(z)|t|^2,$$

where $a(z) := \|\sigma(z)\|_h^2$. Then we set $t = re^{\sqrt{-1}\theta}$ for polar coordinates (r, θ). For a suitable choice of σ, we may assume that $(da)(0) = 0$ and $a(0) = 1$. Hence $\partial\rho = \partial r$ and $\bar{\partial}\rho = \bar{\partial}r$ at the point p. Then

$$\sqrt{-1}\,\partial x \wedge \bar{\partial}x \; = \; \sqrt{-1}\,\frac{dt \wedge d\bar{t}}{|t|^2} \; = \; 2\,\frac{dr \wedge d\theta}{r} \; = \; dx \wedge d\theta \qquad (9.18)$$

at p. For simplicity, put $q := \lambda'$. Then the Jacobian for the mapping $q : \mathbb{R} \to [-1, 1]$, $x \mapsto q(x)$, is $\lambda''(x)$. In view of (9.15) and (9.18), we obtain

$$\int_X \lambda'(x)^\alpha \omega^n \; = \; c_0 b_\alpha, \qquad \alpha = 0, 1, 2, \qquad (9.19)$$

where $c_0 := n \int_N \phi^k$. Now by (9.17), $\int_X \{C_1 + C_2\lambda'(x)\}\omega^n = \int_X e^{f_\omega}\omega^n = \int_X \omega^n$. This together with (9.19) allows us to obtain

$$C_1 b_0 + C_2 b_1 \; = \; b_0. \qquad (9.20)$$

On the other hand, by (9.17) and $\lambda'(x) \in \mathrm{Ker}_{\mathbb{R}}(D_\omega + 1)$, we have $\int_X \lambda'(x)\{C_1 + C_2\lambda'(x)\}\omega^n = \int_X \lambda'(x)e^{f_\omega}\omega^n = -\int_X\{D_\omega\lambda'(x)\}e^{f_\omega}\omega^n = 0$. Hence by (9.19),

$$C_1 b_1 + C_2 b_2 \; = \; 0. \qquad (9.21)$$

If $b_1 = 0$, then by (9.20) and (9.21), $(C_1, C_2) = (1, 0)$. Hence by the vanishing $b_1 = 0$ of the Futaki character, we see from Koiso and Sakane's result [36] that ω is a Kähler–Einstein metric on X, since in this case (9.17) above reduces to the equation for ω to be a Kähler–Einstein metric (see for instance [43]).

Hence we may assume $b_1 \neq 0$. Now by (9.20) and (9.21), we obtain $(C_1, C_2) = (b_0 b_2/(b_0 b_2 - b_1^2), -b_0 b_1/(b_0 b_2 - b_1^2))$. For the time being, assuming that

$$|b_1| < b_2, \qquad (9.22)$$

we shall show that ω is generalized Kähler–Einstein. By $1 \leq \lambda'(x) \leq 1$, the right-hand side of (9.17) is bounded from below by a positive real constant as follows:

$$C_1 + C_2\lambda'(x) \; \geq \; C_1 - |C_2| \; = \; \frac{b_0(b_2 - |b_1|)}{b_0 b_2 - b_1^2} \; > \; 0.$$

Put $A(s) := -\int_{-1}^s q(C_1 + C_2 q)\prod_{i=1}^k(1 - \beta_i q)\,dq$, $s \in \mathbb{R}$. By (9.21), $A(1) = A(-1) = 0$, where the order of zeroes of $A(s)$ at $s = \pm 1$ is 1 (cf. [43]). Then $0 < A(s) \leq A(0)$ and $A'(s)/s < 0$ for $0 \neq |s| < 1$. Hence $-A'(s)/(s\,A(s))$ is a positive rational function free from poles and zeroes on the open interval $(-1, 1)$, and by (9.22), has poles of order 1 at $s = \pm 1$. It then follows that the function $x : (-1, 1) \to \mathbb{R}$ defined by

$$x(s) := -\int_0^s A'(q)/(q\, A(q))\, dq$$

is monotone increasing, and maps $(-1, 1)$ diffeomorphically onto $\mathbb{R}$, since in a neighborhood of $s = 1$ (resp. $s = -1$), $x(s)$ is expressible as $-\log(1 - s) +$ real analytic function (resp. $\log(1 + s) +$ real analytic function). Let $s = s(x) : \mathbb{R} \to (-1, 1)$ be the inverse function of $x = x(s) : (-1, 1) \to \mathbb{R}$ above. Let us now consider the function $\lambda = \lambda(x)$ of x defined by $\lambda(x) := -\log(A(s))$. Then

$$\lambda'(x) = -\frac{A'(s)}{A(s)} \cdot \frac{ds}{dx} = \frac{A'(s)}{A(s)} \cdot \frac{sA(s)}{A'(s)} = s.$$

In particular, we obtain the equality $\lambda''(x) = s'(x) = 1/x'(s) = -sA(s)/A'(s)$, where on the right-hand side, the numerator $-s\, A(s)$ is $-s\, e^{-\lambda(x)}$, and the denominator $A'(s)$ is $-s\, (C_1 + C_2 s) \prod_{i=1}^{k}(1 - \beta_i s)$. Hence by $s = \lambda'(x)$,

$$e^{-\lambda(x)} \left\{ \lambda''(x) \prod_{i=1}^{k}(1 - \beta_i \lambda'(x)) \right\}^{-1} = C_1 + C_2 \lambda'(x),$$

i.e., for the function $s = s(x)$ above, $\lambda(x) = -\log(A(s))$ satisfies (9.17). Since $e^{-\lambda(x)} = A(s)$ has zeroes of order 1 at $s = \pm 1$, and since the real analytic function $\lambda''(x) = -sA(s)/A'(s)$ of $s = s(x)$ is non-vanishing on the interval $(-1, 1)$ with zeroes of order 1 at $s = \pm 1$, we see from (9.18) that Ω extends to a volume form on X, and ω extends to a Kähler form on X (cf. [43]). Thus we obtain:

Theorem 9.4 *If $|b_1| < b_2$, then X in (9.12) admits a generalized Kähler–Einstein metric, and in particular $\gamma_X < 1$.*

Remark 9.1 Actually, by using the generalized Kähler–Einstein metric ω, we see from (9.16) and (9.17) the following:

$$\gamma_X = \max_X \mu_\omega = \max_X(1 - e^{f_\omega}) = \max_X \left\{ (1 - C_1) - C_2 \lambda'(x) \right\}$$

$$= 1 - C_1 + |C_2| = \frac{b_0 |b_1| - b_1^2}{b_0 b_2 - b_1^2} < 1.$$

Finally, we consider the case $|b_1| \geq b_2$. Let Ω in (9.14) be chosen in such a way that it extends to a volume form on X and that the associated ω extends to a Kähler form on X. Here $\lambda(x)$ is not necessarily assumed to satisfy (9.17). Since every holomorphic automorphism of X induces a holomorphic automorphism of N, and since every infinitesimal holomorphic action on N lifts to an infinitesimal holomorphic action on L, we can write $\mathfrak{g} = H^0(X, \mathscr{O}(TX))$ as a semidirect sum

$$\mathfrak{g} = H^0(N, \mathscr{O}(TN)) \ltimes \mathfrak{h},$$

where $\mathfrak{h}$ is the Lie algebra of all holomorphic automorphisms of X preserving the fibers of X over N. Since the $\mathbb{C}^*$-action on L commutes with any infinitesimal holomorphic action on N, the holomorphic vector field $u := \mathrm{grad}_\omega^{\mathbb{C}} \lambda'(x) \in \mathfrak{k}$ on X associated to the $\mathbb{C}^*$-action on L commutes with the Lie algebra $H^0(N, \mathscr{O}(TN))$. Let K_0 be the identity component of the group of isometries of (N, ϕ). Since (N, ϕ) is Kähler–Einstein, $H^0(N, \mathscr{O}(TN))$ is a complexification $\mathfrak{k}_0^{\mathbb{C}}$ of the Lie algebra $\mathfrak{k}_0 := \mathrm{Lie}(K_0)$ of K_0 such that

$$\mathfrak{k} = \mathfrak{k}_0 \oplus \mathbb{R}u,$$

where K_0 is assumed to fix the Hermitian metric h for L. Since the extremal vector field y_ω on X is in the center of $\mathfrak{k}^{\mathbb{C}}$, by (9.1), we can write y_ω as a sum $w + \alpha_1 u$ for some $w \in \mathfrak{k}_0$ and a real constant α_1. Since $w = \mathrm{grad}_\omega^{\mathbb{C}} \varphi$ for some $\varphi \in \mathfrak{k}_\omega$, we have

$$\mathrm{pr}(1 - e^{f_\omega}) = \varphi + \alpha_1 \lambda'(x) + \alpha_2$$

for some real constant α_2. For $\alpha_0 := \{\int_X \lambda'(x)\varphi\,\omega^n\}(b_0/c_0)/(b_0 b_2 - b_1^2)$, we put $\varphi_0 := \varphi - \alpha_0\{\lambda'(x) - (b_1/b_0)\}$, $C_3 := \alpha_2 - \alpha_0(b_1/b_0)$ and $C_4 := \alpha_0 + \alpha_1$. Then

$$\mathrm{pr}(1 - e^{f_\omega}) = \varphi_0 + C_4 \lambda'(x) + C_3. \tag{9.23}$$

Since $\lambda'(x) - (b_1/b_0) \in \mathfrak{k}_\omega$, we have $\varphi_0 \in \mathfrak{k}_\omega$. Then by (9.19),

$$\int_X \lambda'(x)\varphi_0\,\omega^n = \int_X \lambda'(x)\varphi\,\omega^n - \alpha_0 \left\{ \int_X \lambda'(x)^2 \omega^n - (b_1/b_0)\int_X \lambda'(x)\omega^n \right\}$$

$$= \left\{ \int_X \lambda'(x)\varphi\,\omega^n \right\} \left(1 - b_0 \cdot \frac{b_2 - (b_1/b_0)b_1}{b_0 b_2 - b_1^2} \right) = 0.$$

For $\tilde{\mathfrak{k}}_\omega := \mathfrak{k}_\omega \oplus \mathbb{R}$, let $\tilde{\mathfrak{k}}_\omega^\perp := \left\{ \varphi \in \mathfrak{k}_\omega^\perp \,;\, \int_X \varphi\omega^n = 0 \right\}$ be its orthogonal complement in $L^2(X, \omega)_\mathbb{R}$. Then for the natural projection $\tilde{\mathrm{pr}} : L^2(X, \omega)_\mathbb{R} (= \tilde{\mathfrak{k}}_\omega \oplus \tilde{\mathfrak{k}}_\omega^\perp) \to \tilde{\mathfrak{k}}_\omega$, we see from $\int_X (1 - e^{f_\omega})\omega^n = 0$ that

$$\mathrm{pr}(1 - e^{f_\omega}) = \tilde{\mathrm{pr}}(1 - e^{f_\omega}) = 1 - \tilde{\mathrm{pr}}(e^{f_\omega}). \tag{9.24}$$

Now by (9.23) and (9.24), $\widetilde{\mathrm{pr}}(e^{f_\omega}) = -\varphi_0 - C_4\lambda'(x) + (1 - C_3)$, and we write e^{f_ω} as a sum $\widetilde{\mathrm{pr}}(e^{f_\omega}) + \zeta$ for some $\zeta \in \tilde{\mathfrak{k}}_\omega^\perp$. By $\lambda'(x) \in \mathrm{Ker}_\mathbb{R}(D_\omega + 1) \subset \tilde{\mathfrak{k}}_\omega$ and (9.19),

$$
\begin{aligned}
\{-C_4 b_2 + (1 - C_3) b_1\} c_0 &= \int_X \lambda'(x)\{-\varphi_0 - C_4\lambda'(x) + (1 - C_3)\}\omega^n \\
&= \int_X \lambda'(x)\,\widetilde{\mathrm{pr}}(e^{f_\omega})\,\omega^n = \int_X \lambda'(x)\,(e^{f_\omega} - \zeta)\omega^n \\
&= \int_X \lambda'(x)\,e^{f_\omega}\,\omega^n = \int_X -\{D_\omega\lambda'(x)\}\,e^{f_\omega}\,\omega^n = 0.
\end{aligned}
$$

Hence $C_4 b_2 = (1 - C_3) b_1$. On the other hand, by (9.19) and $\varphi_0 \in \mathfrak{k}_\omega$,

$$
\begin{aligned}
\{-C_4 b_1 + (1 - C_3) b_0\} c_0 &= \int_X \{-\varphi_0 - C_4\lambda'(x) + (1 - C_3)\}\omega^n \\
&= \int_X \widetilde{\mathrm{pr}}(e^{f_\omega})\omega^n = \int_X (e^{f_\omega} - \zeta) \cdot 1 \cdot \omega^n = \int_X e^{f_\omega}\omega^n = \int_X \omega^n = b_0 c_0,
\end{aligned}
$$

and hence $-C_4 b_1 + (1 - C_3) b_0 = b_0$. From these, we obtain

$$
(C_3, C_4) = (-b_1^2/(b_0 b_2 - b_1^2),\ b_0 b_1/(b_0 b_2 - b_1^2)).
$$

This together with (9.24) and $\widetilde{\mathrm{pr}}(e^{f_\omega}) = -\varphi_0 - C_4\lambda'(x) + (1 - C_3)$ implies that

$$
\mathrm{pr}(1 - e^{f_\omega}) = \varphi_0 + C_3 + C_4\lambda'(x) = \varphi_0 + \frac{b_0 b_1 \lambda'(x) - b_1^2}{b_0 b_2 - b_1^2}. \tag{9.25}
$$

Let us now assume that $\mathrm{Aut}^0(N)$ is semisimple. For instance, such a condition is satisfied if N is a rational homogeneous space or if $\mathrm{Aut}(N)$ is discrete. For $\lambda' = \lambda'(x) \in \mathrm{Ker}_\mathbb{R}(D_\omega + 1)$ above, by setting $\mathfrak{g}_N := H^0(N, \mathcal{O}(TN))$, we consider

$$
\hat{\mathfrak{g}}_\omega := \left\{ \phi \in \mathfrak{g}_\omega ;\ \int_X \phi\lambda'\omega^n = 0 \ \text{ and } \ \mathrm{grad}_\omega^\mathbb{C}\phi \in \mathfrak{g}_N \oplus \mathbb{C}u \right\},
$$

which is a Lie subalgebra of $\mathfrak{g}_\omega$ by the Poisson bracket $[\ ,\ ]_\omega$ in terms of ω as follows: Let $\phi_1, \phi_2 \in \hat{\mathfrak{g}}_\omega$. Since u is in the center of $\mathfrak{g}_N \oplus \mathbb{C}u$, we obtain $[\phi_2, \lambda']_\omega = 0$. Hence $\int_X [\phi_1, \phi_2]_\omega \lambda'\omega^n = \int_X \phi_1 [\phi_2, \lambda']_\omega\omega^n = 0$. This means that $[\phi_1, \phi_2]_\omega \in \hat{\mathfrak{g}}_\omega$. For the Lie subalgebra $\hat{\mathfrak{g}} := \{\mathrm{grad}_\omega^\mathbb{C}\phi\ ; \ \phi \in \hat{\mathfrak{g}}_\omega\}$ of $\mathfrak{g}$, we have a Lie algebra isomorphism

$$
\mathrm{pr}_1 :\ \hat{\mathfrak{g}} \cong \mathfrak{g}_N,
$$

where pr_1 is the restriction to $\hat{\mathfrak{g}}$ of the projection: $\mathfrak{g}_N \oplus \mathbb{C}u \to \mathfrak{g}_N$ to the first factor. By the semisimplicity of $\mathfrak{g}_N$, we have $[\mathfrak{g}_N, \mathfrak{g}_N] = \mathfrak{g}_N$, and hence $[\hat{\mathfrak{g}}, \hat{\mathfrak{g}}] = \hat{\mathfrak{g}}$. In particular, for the Futaki character $\mathscr{F} : \mathfrak{g} \to \mathbb{C}$, we have

$$\mathscr{F}(\hat{\mathfrak{g}}) = \mathscr{F}([\hat{\mathfrak{g}}, \hat{\mathfrak{g}}]) = 0.$$

Since φ_0 in (9.23) belongs to $\hat{\mathfrak{g}}_\omega$, we have $\mathrm{grad}_\omega^{\mathbb{C}} \varphi_0 \in \hat{\mathfrak{g}} \subset \mathrm{Ker}\, \mathscr{F}$. Then by (9.23), since $\varphi_0 \in \mathfrak{k}_\omega$, and since $\int_X \varphi_0 \lambda' \omega^n = 0$, it follows that

$$0 = \mathscr{F}(\mathrm{grad}_\omega^{\mathbb{C}} \varphi_0) = \int_X \varphi_0 \,\mathrm{pr}(1 - e^{f_\omega})\omega^n$$

$$= \int_X \varphi_0(\varphi_0 + C_4\lambda' + C_3)\omega^n = \int_X \varphi_0^2 \omega^n,$$

and hence $\varphi_0 = 0$. Substituting this into (9.25), we obtain $\mathrm{pr}(1 - e^{f_\omega}) = \{-b_1^2 + b_0 b_1 \lambda'(x)\}/(b_0 b_2 - b_1^2)$. Note that $\lambda'_{|D_0} = -1$ and $\lambda'_{|D_\infty} = 1$. Hence

$$\gamma_X = \max_X \mathrm{pr}(1 - e^{f_\omega}) = \frac{b_0|b_1| - b_1^2}{b_0 b_2 - b_1^2} \geq 1, \qquad (9.26)$$

where for the last inequality, the condition $|b_1| \geq b_2$ is used. Then by Theorem 9.3, X admits no generalized Kähler–Einstein metrics. By summing up, we obtain:

Theorem 9.5 *For X in (9.12), assume that $\mathrm{Aut}^0(N)$ is semisimple. Assume further that K_0 fixes the Hermitian metric h for L. Then*

$$\gamma_X = \frac{b_0|b_1| - b_1^2}{b_0 b_2 - b_1^2}.$$

Hence in this case $\gamma_X < 1$ if and only if $|b_1| < b_2$, and furthermore X admits a generalized Kähler–Einstein metric if and only if $\gamma_X < 1$.

Example 9.1 (cf. [45]) Let $N = \mathbb{P}^k(\mathbb{C})$ and $L = \mathscr{O}_{\mathbb{P}^k}(1)$, $k \geq 1$. By $c_1(N) = L^{\otimes k+1}$, we have $b_\alpha = \int_{-1}^1 y^\alpha \{1 - (k+1)^{-1}y\}^k dy$. Note that $b_1 < 0$. It is easy to check that

$$b_2 - |b_1| = b_2 + b_1 \geq \frac{2}{3} \cdot \frac{1}{k+1} > 0.$$

Hence $X := \mathbb{P}(\mathscr{O}_{\mathbb{P}^k} \oplus \mathscr{O}_{\mathbb{P}^k}(1))$ admits a generalized Kähler–Einstein metric, where by $b_1 \neq 0$, X admits no Kähler–Einstein metrics. For instance, if $k = 1$, then X is the Hirzebruch surface F_1, and by $(b_0, b_1, b_2) = (2, -1/3, 2/3)$ and Remark 9.1,

$$\gamma_X = \frac{b_0|b_1| - b_1^2}{b_0 b_2 - b_1^2} = \frac{5}{11} < 1.$$

Example 9.2 (cf. [45]) Let $N = \mathbb{P}^2(\mathbb{C})$ and $L = \mathscr{O}_{\mathbb{P}^2}(2)$. Since $b_\alpha = \int_{-1}^{1} y^\alpha \{1 - (2y/3)\}^2 dy$, we have $(b_0, b_1, b_2) = (62/27, -8/9, 38/45)$. Then

$$|b_1| = \frac{8}{9} > \frac{38}{45} = b_2.$$

Hence $X := \mathbb{P}(\mathscr{O}_{\mathbb{P}^2} \oplus \mathscr{O}_{\mathbb{P}^2}(2))$ admits no generalized Kähler–Einstein metrics. Moreover by (9.26), we have the inequality

$$\gamma_X \geq \frac{b_0|b_1| - b_1^2}{b_0 b_2 - b_1^2} = \frac{380}{349} > 1.$$

However, it is also known (cf. [27, 33]) that X admits an extremal Kähler metric in the class $c_1(M)$. This last fact was pointed out to the author by S. Nakamura.

9.6 Extremal Metrics on Generalized Kähler–Einstein Manifolds

In this section, we briefly discuss a fact concerning the existence of extremal Kähler metrics on generalized Kähler–Einstein manifolds. The proof down below is due to S. Nakamura. For an n-dimensional Fano manifold X, let $\mathscr{K}$ be its Kähler class $c_1(X)$. By fixing a reference metric ω_0 in $\mathscr{K}$, we can write

$$\mathscr{K} = \{\omega_\varphi ; \varphi \in C^\infty(X)_{\mathbb{R}} \text{ is such that } \omega_\varphi \text{ is Kähler}\},$$

where $\omega_\varphi := \omega_0 + dd^c\varphi$. Let $\mu_\omega := \mathrm{pr}(1 - e^{f_\omega}) = \mathrm{pr}(S_\omega - n)$ be as in Sect. 9.4. Put $V := \int_X \omega_0^n = c_1(X)^n[X]$. We now consider Aubin's functional $J : \mathscr{K} \to \mathbb{R}_{\geq 0}$, Ding's functional $D : \mathscr{K} \to \mathbb{R}$, and the K-energy $\kappa : \mathscr{K} \to \mathbb{R}$ defined by

$$J(\varphi) := \frac{1}{V} \int_0^1 \left\{ \int_X \dot{\varphi}_s (\omega_0^n - \omega_s^n) \right\} ds,$$

$$D(\varphi) := -\frac{1}{V} \int_0^1 \left(\int_X \dot{\varphi}_s \omega_s^n \right) ds - \log \left(\frac{1}{V} \int_X e^{f_0 - \varphi} \omega_0^n \right).$$

$$\kappa(\varphi) := -\frac{1}{V} \int_0^1 \left\{ \int_X \dot{\varphi}_s (S_{\omega_s} - n) \omega_s^n \right\} ds,$$

where $\{\varphi_s\}_{0 \leq s \leq 1}$ is a piecewise smooth path in $C^\infty(X)_{\mathbb{R}}$ as in Sect. 6.3 such that $\varphi_0 = 0$ and $\varphi_1 = \varphi$. Here by $\dot{\varphi}_s$, we mean $\partial \varphi_s / \partial s$, while ω_{φ_s} and $f_{\omega_{\varphi_s}}$ are written as ω_s and f_s, respectively. The corresponding modified energies are

$$\hat{J}(\varphi) := \frac{1}{V} \int_0^1 \left\{ \int_X \dot{\varphi}_s ((1 - \mu_{\omega_0})\omega_0^n - (1 - \mu_{\omega_s})\omega_s^n) \right\} ds, \qquad (9.27)$$

$$\hat{D}(\varphi) := -\frac{1}{V} \int_0^1 \left\{ \int_X \dot{\varphi}_s (1 - \mu_{\omega_s})\omega_s^n \right\} ds - \log(\frac{1}{V} \int_X e^{f_0 - \varphi}\omega_0^n), \qquad (9.28)$$

$$\hat{\kappa}(\varphi) := -\frac{1}{V} \int_0^1 \left\{ \int_X \dot{\varphi}_s (S_{\omega_s} - n - \mu_{\omega_s})\omega_s^n \right\} ds. \qquad (9.29)$$

First by (9.29), ω is a critical point of $\hat{\kappa}$ if and only if $S_{\omega_s} - n = \mu_{\omega_s}$, i.e., ω is an extremal Kähler metric. Next by differentiating (9.27), we obtain

$$\frac{d}{ds} \hat{D}(\varphi_s) = -\frac{1}{V} \int_X \dot{\varphi}_s (1 - \mu_{\omega_s})\omega_s^n + \frac{\int_X e^{f_0 - \varphi_s} \dot{\varphi}_s \, \omega_0^n}{\int_X e^{f_0 - \varphi_s} \, \omega_0^n}. \qquad (9.30)$$

For the time being, we assume the following:

Claim $\quad f_s = f_0 - \varphi_s + \log(\omega_0^n / \omega_s^n) + \log \left(V / \int_X e^{f_0 - \varphi_s} \omega_0^n \right).$

By this claim, we immediately see that $e^{f_0 - \varphi_s} = e^{f_s}(\omega_s^n / \omega_0^n)(\int_X e^{f_0 - \varphi_s}\omega_0^n / V)$. Substituting this into (9.30), we obtain

$$\frac{d}{ds} \hat{D}(\varphi_s) = -\frac{1}{V} \int_X \dot{\varphi}_s (1 - \mu_{\omega_s})\omega_s^n + \frac{1}{V} \int_X e^{f_s} \dot{\varphi}_s \, \omega_s^n = -\frac{1}{V} \int_X \dot{\varphi}_s (1 - e^{f_s} - \mu_{\omega_s})\omega_s^n.$$

Hence ω is a critical point of $\hat{D}$ if and only if $1 - e^{f_\omega} = \mu_\omega$, i.e., ω is a generalized Kähler–Einstein metric. We shall now prove the above claim.

Proof Note that $dd^c f_s = \text{Ric}(\omega_s) - \omega_s = \{\text{Ric}(\omega_0) - \omega_0\} - dd^c\varphi_s + \{\text{Ric}(\omega_s) - \text{Ric}(\omega_0)\} = dd^c\{f_0 - \varphi_s + \log(\omega_0^n / \omega_s^n)\}$. Hence $f_s = f_0 - \varphi_s + \log(\omega_0^n / \omega_s^n) + C$ for some real constant C. Then Claim follows from

$$V = \int_X e^{f_s}\omega_s^n = \int_X e^{f_0 - \varphi_s + \log(\omega_0^n / \omega_s^n) + C}\omega_s^n = e^C \int_X e^{f_0 - \varphi_s}\omega_0^n.$$

$\square$

Note that the derivative of $\log \omega_s^n$ with respect to s is $\Delta_{\omega_s} \dot{\varphi}_s$. Then by differentiating the equality in Claim with respect to s, we have

$$\dot{f}_s = -\dot{\varphi}_s - \Delta_{\omega_s}\dot{\varphi}_s + \frac{\int_X e^{f_0 - \varphi_s} \dot{\varphi}_s \, \omega_0^n}{\int_X e^{f_0 - \varphi_s} \, \omega_0^n}.$$

We integrate this over X by the volume form $(1/V)\omega_s^n$. It then follows that

$$\frac{1}{V}\int_X \dot{f_s}\omega_s^n = -\frac{1}{V}\int_X \dot{\varphi}_s\omega_s^n + \frac{\int_X e^{f_0-\varphi_s}\dot{\varphi}_s\,\omega_0^n}{\int_X e^{f_0-\varphi_s}\,\omega_0^n}.$$

Now by comparing this with (9.30), we obtain

$$\frac{d}{ds}\hat{D}(\varphi_s) = \frac{1}{V}\int_X \dot{f_s}\omega_s^n + \frac{1}{V}\int_X \dot{\varphi}_s\mu_{\omega_s}\omega_s^n. \tag{9.31}$$

Since $(d/ds)(\int_X f_s\omega_s^n) - \int_X \dot{f_s}\omega_s^n = \int_X f_s(\Delta_{\omega_s}\dot{\varphi}_s)\omega_s^n = \int_X \dot{\varphi}_s(\Delta_{\omega_s}f_s)\omega_s^n$, and since $S_{\omega_s} - n = \Delta_{\omega_s}f_s$, we see from (9.29) and (9.31) the following:

$$
\begin{aligned}
\frac{d}{ds}\hat{\kappa}(\varphi_s) &= -\frac{1}{V}\int_X \dot{\varphi}_s(S_{\omega_s} - n - \mu_{\omega_s})\omega_s^n = -\frac{1}{V}\int_X \dot{\varphi}_s(\Delta_{\omega_s}f_s - \mu_{\omega_s})\,\omega_s^n \\
&= -\frac{d}{ds}\left(\frac{1}{V}\int_X f_s\omega_s^n\right) + \frac{1}{V}\int_X \dot{f_s}\omega_s^n + \frac{1}{V}\int_X \dot{\varphi}_s\mu_{\omega_s}\omega_s^n \\
&= -\frac{d}{ds}\left(\frac{1}{V}\int_X f_s\omega_s^n\right) + \frac{d}{ds}\hat{D}(\varphi_s).
\end{aligned}
$$

Integrate this equality over the interval $[0, 1]$. Then in view of $\varphi_0 = 0$ and $\varphi_1 = \varphi$, since $\hat{\kappa}(0) = 0$ and $\hat{D}(0) = 0$, we obtain

$$\hat{\kappa}(\varphi) = -\frac{1}{V}\int_X f_{\omega_\varphi}\omega_\varphi^n + \frac{1}{V}\int_X f_{\omega_0}\omega_0^n + \hat{D}(\varphi) \tag{9.32}$$

for all ω_φ in $\mathscr{K}$. Since $x + 1 \le e^x$ for all real numbers x, we have

$$\int_X (f_{\omega_\varphi} + 1)\omega_\varphi^n \le \int_X e^{f_{\omega_\varphi}}\omega_\varphi^n = \int_X \omega_\varphi^n,$$

so that $\int_X f_{\omega_\varphi}\omega_\varphi^n \le 0$ for all ω_φ in $\mathscr{K}$. Hence by setting $C_0 := -(1/V)\int_X f_{\omega_0}\omega_0^n$, it follows from (9.32) that

$$\hat{\kappa}(\varphi) \ge \hat{D}(\varphi) - C_0 \tag{9.33}$$

for all ω_φ in $\mathscr{K}$, where C_0 is a nonnegative real constant independent of φ. Let ω be a generalized Kähler–Einstein metric. Then the following cases are possible:

Case 1: $y_\omega = 0$. Then $1 - e^{f_\omega} = \mu_\omega = 0$. Hence ω is Kähler–Einstein, and in particular extremal Kähler.

Case 2: $y_\omega \neq 0$. In this case, by taking the circle group $T_c := \exp(\mathbb{R}y_\omega^\mathbb{R})$, we choose a T_c-invariant Kähler metric in $\mathscr{K}$ as the reference metric ω_0 above. Let G_0 be the centralizer of T_c in $G = \mathrm{Aut}^0(X)$. Put

$$\mathscr{H}^{T_c} = \{\varphi \in C^\infty(X)_\mathbb{R} \, ; \, \omega_\varphi \in \mathscr{K} \text{ and } \varphi \text{ is } T_c \text{ -invariant}\}.$$

Since ω is generalized Kähler–Einstein, we have $\gamma_X < 1$. Recall the following theorem of Li and Zhou [41]:

Theorem 9.6

(1) If $\gamma_X < 1$, then there exist positive constants C_1 and C_2 independent of φ such that $0 \leq C_1 J(\varphi) \leq \hat{J}(\varphi) \leq C_2 J(\varphi)$ for all $\varphi \in \mathscr{H}^{T_c}$.

(2) If X admits a generalized Kähler–Einstein metric, then there exist positive constants C_3 and C_4 independent of φ such that $\hat{D}(\varphi) \geq C_3 \inf_{g \in G_0} \hat{J}(\varphi_g) - C_4$ for all $\varphi \in \mathscr{H}^{T_c}$, where $\varphi_g \in \mathscr{H}^{T_c}$ is such that $g^*(\omega_\varphi) = \omega_0 + dd^c \varphi_g$ and that $\int_X \varphi_g \omega_0^n = 0$.

Since ω is generalized Kähler–Einstein, by (2) of Theorem 9.6, there exist positive constants C_3 and C_4 as above such that

$$\hat{D}(\varphi) \geq C_3 \inf_{g \in G_0} \hat{J}(\varphi_g) - C_4, \qquad \varphi \in \mathscr{H}^{T_c}.$$

This together with (9.33) allows us to obtain

$$\hat{\kappa}(\varphi) \geq \hat{D}(\varphi) - C_0 \geq C_3 \inf_{g \in G_0} \hat{J}(\varphi_g) - (C_4 + C_0). \tag{9.34}$$

Hence the functional $\hat{\kappa}(\varphi)$, $\varphi \in \mathscr{H}^{T_c}$, is bounded from below by a real constant independent of the choice of φ, since in (9.34), we see from (1) of Theorem 9.6 that

$$\inf_{g \in G_0} \hat{J}(\varphi_g) \geq C_1 \inf_{g \in G_0} J(\varphi_g) \geq 0. \tag{9.35}$$

We now apply the following theorem of He [29, Theorem 2]. Here, we adapt his result to the anticanonical class for Fano manifolds.

Theorem 9.7 *The Kähler class $\mathscr{K}$ on X admits an extremal Kähler metric if the following conditions are satisfied:*

(1) $\hat{\kappa}(\varphi)$, $\varphi \in \mathscr{H}^{T_c}$, is bounded from below.

(2) For every sequence φ_i, $i=1,2,\ldots$, in $\mathscr{H}^{T_c}$ such that $\inf_{g \in G_0} J((\varphi_i)_g) \to +\infty$ as $i \to \infty$, we always have $\hat{\kappa}(\varphi_i) \to +\infty$ as $i \to \infty$.

Remark 9.2 The modified K-energy is called *proper modulo the action of G_0* if the above conditions (1) and (2) are satisfied.

As we saw above, by (9.34) and (9.35), condition (1) of Theorem 9.7 is satisfied. For condition (2), again by (9.34) and (9.35), we see that

$$\hat{\kappa}(\varphi_i) \geq C_3 C_1 \inf_{g \in G_0} J((\varphi_i)_g) - (C_4 + C_0).$$

Since $\inf_{g \in G_0} J((\varphi_i)_g) \to +\infty$ as $i \to \infty$, it then follows that $\hat{\kappa}(\varphi_i) \to +\infty$ as $i \to \infty$. Hence (2) is also satisfied. Now by applying Theorem 9.7, the Kähler class $\mathscr{K}$ on X admits an extremal Kähler metric. By summing up, we obtain:

Theorem 9.8 *If a Fano manifold X admits a generalized Kähler–Einstein metric, then the Kähler class $c_1(X)$ admits an extremal Kähler metric.*

9.7   The Product Formula for the Invariant γ_X

The results in this section are due to Y. Nitta and S. Saito. For the product $X = X_1 \times X_2$ of Fano manifolds X_1 and X_2, we have the following formula for γ_X:

Theorem 9.9 $\gamma_X = \gamma_{X_1} + \gamma_{X_2}$.

Proof For $i = 1, 2$, we choose a Kähler form ω_i on X_i in the class $c_1(X_i)$, and let $\pi_i : X (= X_1 \times X_2) \to X_i, i = 1, 2$, be the projection to the i-th factor. Put

$$\omega := \pi_1^* \omega_1 + \pi_2^* \omega_2, \qquad (9.36)$$

where we also put $n_1 := \dim X_1$ and $n_2 := \dim X_2$. Hence for $n := n_1 + n_2$, we have $\omega^n = \{n!/(n_1! n_2!)\}(\pi_1^* \omega_1^n)(\pi_2^* \omega_2^n)$, so that $\mathrm{Ric}(\omega) = -dd^c \log \omega^n$ is expressible as

$$\mathrm{Ric}(\omega) = -\pi_1^* dd^c \log \omega_1^n - \pi_2^* dd^c \log \omega_2^n = \pi_1^* \mathrm{Ric}(\omega_1) + \pi_2^* \mathrm{Ric}(\omega_2). \qquad (9.37)$$

Note that $S_\omega = \mathrm{Tr}_\omega \mathrm{Ric}(\omega)$ and $S_{\omega_i} = \mathrm{Tr}_{\omega_i} \mathrm{Ric}(\omega_i), i = 1, 2$. Hence by comparing the equalities (9.36) and (9.37), we obtain

$$S_\omega = \pi_1^* S_{\omega_1} + \pi_2^* S_{\omega_2}. \qquad (9.38)$$

Let pr : $L^2(X, \omega) \to \mathfrak{k}_\omega$ be the orthogonal projection as in Sect. 9.2. Similarly, we consider the orthogonal projections $\mathrm{pr}_i : L^2(X_i, \omega_i) \to \mathfrak{k}_{\omega_i}, i = 1, 2$. Then by (9.36), $\mathfrak{k}_\omega = \pi_1^* \mathfrak{k}_{\omega_1} \oplus \pi_2^* \mathfrak{k}_{\omega_2}$. It then follows from (9.38) that

$$\mathrm{pr}(S_\omega) = \mathrm{pr}(\pi_1^* S_{\omega_1}) + \mathrm{pr}(\pi_2^* S_{\omega_2}) = \pi_1^*(\mathrm{pr}_1 S_{\omega_1}) + \pi_2^*(\mathrm{pr}_2 S_{\omega_2}). \qquad (9.39)$$

Let a point $x = (x_1, x_2)$ run through the set $X = X_1 \times X_2$. In view of the equalities $\mathrm{pr}(S_\omega - n) = \mathrm{pr}(S_\omega)$ and $\mathrm{pr}_i(S_{\omega_i} - n_i) = \mathrm{pr}_i(S_{\omega_i})$, $i = 1, 2$, we obtain

$$\gamma_X = \max_{x \in X} \mathrm{pr}(S_\omega)(x) = \max_{x \in X} \{\pi_1^*(\mathrm{pr}_1 S_{\omega_1})(x) + \pi_2^*(\mathrm{pr}_2 S_{\omega_2})(x)\}$$

$$= \max_{x_1 \in X_1}(\mathrm{pr}_1 S_{\omega_1})(x_1) + \max_{x_2 \in X_2}(\mathrm{pr}_2 S_{\omega_2})(x_2) = \gamma_{X_1} + \gamma_{X_2}. \qquad \square$$

From (9.38) and $n = n_1 + n_2$, we obtain the equality $S_\omega - n = \pi_1^*(S_{\omega_1} - n_1) + \pi_2^*(S_{\omega_2} - n_2)$, while by (9.39), $\mathrm{pr}(S_\omega - n) = \pi_1^*(\mathrm{pr}_1(S_{\omega_1} - n_1)) + \pi_2^*(\mathrm{pr}_2(S_{\omega_2} - n_2))$. Hence if ω_1 is extremal Kähler, and in addition if ω_2 is extremal Kähler, then we have both $S_{\omega_1} - n_1 = \mathrm{pr}_1(S_{\omega_1} - n_1)$ and $S_{\omega_2} - n_2 = \mathrm{pr}_2(S_{\omega_2} - n_2)$, and consequently $S_\omega - n = \mathrm{pr}(S_\omega - n)$, i.e., ω is extremal Kähler. Thus we obtain:

Theorem 9.10 *If ω_1 is an extremal Kähler metric in the class $c_1(X_1)$, and if ω_2 is an extremal Kähler metric in the class $c_1(X_2)$, then ω in (9.36) is an extremal Kähler metric in the class $c_1(X)$ for $X = X_1 \times X_2$.*

However, the same thing is not true for generalized Kähler–Einstein metrics. Suppose that a Fano manifold X_0 admits a generalized Kähler–Einstein metric ω_0 which is not Kähler–Einstein. Then by Theorem 9.3,

$$0 < \gamma_{X_0} < 1.$$

Here we have $\gamma_{X_0} \neq 0$ as follows. For contradiction, assume that $\gamma_{X_0} = 0$. Then the Hamiltonian function $\mu_{\omega_0} = \mathrm{pr}(1 - e^{f_{\omega_0}})$ for the extremal vector field on X_0 satisfies

$$0 = \gamma_{X_0} = \max_X \mu_{\omega_0}.$$

Since ω_0 is generalized Kähler–Einstein, we can write $\mu_{\omega_0} = 1 - e^{f_{\omega_0}}$, which has maximum 0 on X, i.e., $f_{\omega_0} \geq 0$ on X. Then by $\int_X e^{f_{\omega_0}} \omega_0^n = \int \omega_0^n$, we obtain

$$f_{\omega_0} = 0 \quad \text{on } X$$

in contradiction to the fact that ω_0 is not Kähler–Einstein.

We now choose arbitrarily an integer k such that $k\gamma_{X_0} \geq 1$. Let us consider the direct product of k copies of X_0,

$$X := X_0^k = X_0 \times X_0 \times \cdots \times X_0.$$

Then by Theorem 9.9, we have $\gamma_X = k\gamma_{X_0} \geq 1$. It then follows from Theorem 9.3 that X admits no generalized Kähler–Einstein metrics.

Example 9.3 Let X_0 be the Hirzebruch surface $F_1 := \mathbb{P}(\mathscr{O}_{\mathbb{P}^1} \oplus \mathscr{O}_{\mathbb{P}^1}(1))$. This is just the case $k = 1$ in Example 9.1. In view of

$$\gamma_{X_0} = 5/11,$$

X_0 admits a generalized Kähler–Einstein metric. Moreover by Calabi [8] (see also [27, 33]), X_0 admits an extremal Kähler metric in the class $c_1(X_0)$. Let X be the direct product of k copies of X_0. Then the following cases are possible:

Case 1: $k = 2$. In this case $\gamma_X = 10/11 < 1$. Since X is a toric Fano manifold, the result in the next section shows that X admits a generalized Kähler–Einstein metric. By Theorem 9.10, X admits an extremal Kähler metric in the class $c_1(X)$.

Case 2: $k \geq 3$. In this case $\gamma_X = 5k/11 > 1$. Then by Theorem 9.3, X admits no generalized Kähler–Einstein metrics, whereas by Theorem 9.10, X admits an extremal Kähler metric in the class $c_1(X)$.

9.8   Yao's Result for Toric Fano Manifolds

We discuss here Yao's result [90] on the existence of generalized Kähler–Einstein metrics on toric Fano manifolds. In this section, by T, we mean the n-dimensional algebraic torus $(\mathbb{C}^*)^n$ endowed with the multiplicative action of T itself:

$$T \times T \ni (t, t') \mapsto t \cdot t' \in T,$$

where $t \cdot t' := (t_1 t_1', \ldots, t_n t_n')$ for $t = (t_1, \cdots, t_n) \in T$ and $t' = (t_1', \cdots, t_n') \in T$. Recall that a Fano manifold X is called *toric* if X is a T-equivariant compactification of T itself such that T is a Zariski open dense subset of X.

Theorem 9.11 *A toric Fano manifold X admits a generalized Kähler–Einstein metric if and only if $\gamma_X < 1$.*

Proof In view of Theorem 9.3, the proof is reduced to showing that there exists a generalized Kähler–Einstein metric on X provided that $\gamma_X < 1$. For the T-equivariant inclusion: $T = \{(t_1, \ldots, t_n)\} \hookrightarrow X$, we define a real-valued functions x_α on the open subset T of X by $|t_\alpha|^2 = e^{x_\alpha}$, $\alpha = 1, 2, \ldots, n$. Let

$$\Omega := n! \, e^{-\lambda(x)} \prod_{\alpha=1}^{n} \left\{ \frac{\sqrt{-1}}{2\pi} \cdot \frac{dt_\alpha \wedge dt_{\tilde{\alpha}}}{|t_a|^2} \right\} \qquad (9.40)$$

be a volume form on X, where $\lambda = \lambda(x)$ is a function of $x = (x_1, \ldots, x_n)$ such that the right-hand side of (9.40) extends to a volume form on X. Put $q_\alpha := \partial\lambda/\partial x_\alpha$. Then $q(x) := (q_1(x), q_2(x), \ldots, q_n(x))$ extends to a moment map

$$q : X \to \mathbb{R}^n = \{(u_1, \ldots, u_n)\} \qquad (9.41)$$

for the T-action on X such that $q_\alpha = q^* u_\alpha$ for all α, where $u = (u_1, \ldots, u_n)$ is the standard coordinates on $\mathbb{R}^n$. Assuming that $\omega := \mathrm{Ric}\,\Omega$ is Kähler on X, we have

$$\omega = -dd^c \log \Omega = \frac{\sqrt{-1}}{2\pi} \sum_{\alpha,\beta} \left(\frac{\partial^2 \lambda}{\partial x_\alpha \partial x_\beta} \cdot \frac{dt_\alpha \wedge dt_{\bar\beta}}{t_\alpha\, t_{\bar\beta}} \right) \tag{9.42}$$

and $q_\alpha \in \mathrm{Ker}_\mathbb{R}(D_\omega + 1)$, $\alpha = 1, 2, \ldots, n$. Then by the same argument as in obtaining the equality (9.16), we see from (9.40) and (9.42) that

$$e^{f_\omega} = \frac{\Omega}{\omega^n} = e^{-\lambda} \left\{ \det \left(\frac{\partial^2 \lambda}{\partial x_\alpha \partial x_\beta} \right)_{1 \le \alpha, \beta \le n} \right\}^{-1}. \tag{9.43}$$

Hence, in order to show that ω is a generalized Kähler–Einstein metric on X, it suffices to solve the following differential equation in λ:

$$\text{R.H.S. of (9.43)} = C_0 + \sum_{\alpha=1}^n C_\alpha \frac{\partial \lambda}{\partial x_\alpha}, \tag{9.44}$$

for suitable real constants C_α, $\alpha = 0, 1, \ldots, n$, such that Ω extends to a volume form on X and that ω extends to a Kähler form on X.

Note that T is a maximal algebraic torus in $G = \mathrm{Aut}^0(X)$ whose maximal compact subgroup acts isometrically on (X, ω). In particular, the Lie algebra $\mathfrak{t}$ of T is a Lie subalgebra of $\mathfrak{k}^\mathbb{C}$. Since the extremal vector field y_ω is in the center of $\mathfrak{k}^\mathbb{C}$ (cf. [26]), y_ω belongs to $\mathfrak{t}$. Let $\widetilde{\mathrm{pr}} : L^2(X, \omega)_\mathbb{R} \to \tilde{\mathfrak{k}}_\omega$ be as in (9.24). Again by (9.24),

$$\widetilde{\mathrm{pr}}(e^{f_\omega}) = 1 - \mathrm{pr}(1 - e^{f_\omega}) = 1 + \mathrm{pr}(e^{f_\omega}), \tag{9.45}$$

where as remarked above, the holomorphic vector field $y_\omega := -\,\mathrm{grad}^\mathbb{C}_\omega\,\mathrm{pr}(e^{f_\omega})$ is in $\mathfrak{t}$. Hence we have real constants C'_α, $\alpha = 0, 1, \ldots, n$, such that (9.45) is written as

$$\widetilde{\mathrm{pr}}(e^{f_\omega}) = C'_0 + \sum_{\alpha=1}^n C'_\alpha \frac{\partial \lambda}{\partial x_\alpha}. \tag{9.46}$$

If ω is generalized Kähler–Einstein, then $\mathrm{pr}(1 - e^{f_\omega}) = 1 - e^{f_\omega}$, and from this equality together with (9.45), we obtain $\widetilde{\mathrm{pr}}(e^{f_\omega}) = e^{f_\omega}$, and in this special case the real constants C_α, $\alpha = 0, 1, \ldots, n$, in (9.44) above reduce to C'_α, $\alpha = 0, 1, \ldots, n$, respectively, Hence for a general ω, where Ω is as in (9.40), C_α are chosen such that

$$C_\alpha := C'_\alpha, \qquad \alpha = 0, 1, \ldots, n. \tag{9.47}$$

As observed in Remark 9.3 below, these C_α are constants depending only on the image $P := \mathrm{Im}(q)$ of the moment map $q : X \to \mathbb{R}^n$ in (9.41), where P is a convex body in $\mathbb{R}^n$ independent of the choice of ω. In (9.41), we have $\partial\lambda/\partial x_\alpha = q_\alpha = q^* u_\alpha$. Hence by (9.45), (9.46) and (9.47), we obtain

$$\mathrm{pr}(1 - e^{f_\omega}) = 1 - \widetilde{\mathrm{pr}}(e^{f_\omega}) = 1 - q^* \sigma,$$

where we define $\sigma := \sum_{\alpha=0}^n C_\alpha u_\alpha$ by setting $u_0 := 1$. Then $\bar{\sigma} := \max_P \sigma$ and $\underline{\sigma} := \min_P \sigma$ are constants independent of the choice of ω, and we obtain

$$\gamma_X = \max_X \mathrm{pr}(1 - e^{f_\omega}) = 1 - \underline{\sigma} \quad \text{and} \quad \min_X \mathrm{pr}(1 - e^{f_\omega}) = 1 - \bar{\sigma}.$$

From the assumption $\gamma_X < 1$, it follows that $\underline{\sigma} > 0$. Since the right-hand side of (9.44) is written as $q^* \sigma$, we have the following a priori bounds:

$$0 < \underline{\sigma} \leq C_0 + \sum_{\alpha=1}^n C_\alpha \frac{\partial\lambda}{\partial x_\alpha} \leq \bar{\sigma}. \tag{9.48}$$

Hence (9.44) is solved by the continuity method as in Wang and Zhu [83], where their method is applied to our situation by replacing the term of the form $\exp(C_0 + \sum_{\alpha=1}^n C_\alpha \partial\lambda/\partial x_\alpha)$ in [83] by the term $C_0 + \sum_{\alpha=1}^n C_\alpha \partial\lambda/\partial x_\alpha$ above. □

Remark 9.3 The constants C_α, $\alpha = 0, 1, \ldots, n$, in (9.44), (9.46), and (9.47) above are computed as follows: For simplicity, put $q_0 := 1$. Let $H = (h_{\alpha\beta})_{1\leq\alpha,\beta\leq n}$ be the Hessian matrix for the function $\lambda = \lambda(x)$ defined by

$$h_{\alpha\beta} := \partial^2\lambda/\partial x_\alpha \partial x_\beta = \partial q_\alpha/\partial x_\beta.$$

If we view q as a function on $\mathbb{R}^n = \{(x_1, \ldots, x_n)\}$, then $\det H$ is the Jacobian of q. For $\theta_\alpha := \arg(t_\alpha)$, we have $\sqrt{-1} dt_\alpha \wedge dt_{\bar{\alpha}}/|t_\alpha|^2 = dx_\alpha \wedge d\theta_\alpha$ as in (9.18). In view of (9.42) together with (9.46) and (9.47), we obtain

$$b_{00} = \int_P du_1 \wedge \cdots \wedge du_n = (2\pi)^{-n} \int_X (dq_1 \wedge d\theta_1) \wedge \cdots \wedge (dq_n \wedge d\theta_n)$$

$$= (2\pi)^{-n} \int_X (\det H)\,(dx_1 \wedge d\theta_1) \wedge (dx_2 \wedge d\theta_2) \cdots \wedge (dx_n \wedge d\theta_n)$$

$$= \int_X \omega^n = \int_X e^{f_\omega} \omega^n = \int_X \widetilde{\mathrm{pr}}(e^{f_\omega}) \omega^n = \int_X \left(\sum_{\beta=0}^n C_\beta q_\beta\right) \omega^n$$

$$= \int_P \left(\sum_{\beta=0}^n C_\beta u_\beta\right) du_1 \wedge \cdots \wedge du_n = \sum_{\beta=0}^n b_{0\beta} C_\beta,$$

where we set $b_{\alpha\beta} := \int_P u_\alpha u_\beta \, du_1 \wedge \cdots \wedge du_n$ for all $\alpha, \beta \in \{0, 1, \ldots, n\}$. Similarly, by $q_\alpha = \partial\lambda/\partial x_\alpha \in \mathrm{Ker}_{\mathbb{R}}(D_\omega + 1)$, $\alpha = 1, \ldots, n$, we obtain

$$
0 = -\int_X (D_\omega q_\alpha) \, e^{f_\omega} \omega^n = \int_X q_\alpha e^{f_\omega} \omega^n = \int_X q_\alpha \, \widetilde{\mathrm{pr}}(e^{f_\omega}) \omega^n
$$

$$
= \int_X q_\alpha \left(\sum_{\beta=0}^n C_\beta q_\beta \right) \omega^n = \int_P u_\alpha \left(\sum_{\beta=0}^n C_\beta u_\beta \right) du_1 \wedge \cdots \wedge du_n = \sum_{\beta=0}^n b_{\alpha\beta} C_\beta
$$

for $\alpha = 1, \ldots, n$. Since $B = (b_{\alpha\beta})_{0 \le \alpha, \beta \le n}$ is a positive-definite square matrix of order $(n+1)$, we consider its inverse matrix $B^{-1} = (b^{\alpha\beta})$. Then

$$
C_\alpha = b^{\alpha 0} b_{00}, \qquad \alpha = 0, 1, \ldots, n.
$$

9.9 Hisamoto's Result on the Existence Problem

For a Fano manifold X, let G_0 be as in Sect. 9.6. We consider the center Z of G_0. For Aubin's functional $J(\varphi)$ in Sect. 9.6, we put

$$
J_Z(\varphi) := \inf_{g \in Z} J(\varphi_g), \qquad \varphi \in \mathscr{H}^{T_c}. \tag{9.49}
$$

Let $(\mathscr{X}, \mathscr{L})$ be a test configuration for (X, K_X^{-1}), where in this section, the ample anticanonical line bundle $L = K_X^{-1}$ is not necessarily assumed to be very ample. For the modified Ding functional $\hat{D}(\varphi)$ in (9.28) and the functional $J_Z(\varphi)$ in (9.49), we have the non-Archimedean version [3, 4] of the functionals $\hat{D}$ and J_Z:

$$
\hat{D}^{NA}(\mathscr{X}, \mathscr{L}) \in \mathbb{R} \quad \text{and} \quad J_Z^{NA}(\mathscr{X}, \mathscr{L}) \in \mathbb{R}_{\ge 0}.
$$

In the Yau–Tian–Donaldson conjecture, $\hat{D}^{NA}(\mathscr{X}, \mathscr{L})$ corresponds to the Donaldson-Futaki invariant of $(\mathscr{X}, \mathscr{L})$, while $J_Z^{NA}(\mathscr{X}, \mathscr{L})$ corresponds to the asymptotic L^1- norm of the test configuration $(\mathscr{X}, \mathscr{L})$. A recent result of Hisamoto [32] together with Yao's work [91] shows that:

Theorem 9.12 *A Fano manifold X admits a generalized Kähler–Einstein metric if and only if the following conditions are satisfied:*

* *The obstruction of Matsushima's type vanishes, i.e., G_0 is reductive;*
* *(X, L) is uniformly Ding stable relative to Z.*

In the above theorem, (X, L) is called *uniformly Ding stable relative to Z* if there exists a positive real constant $\delta > 0$ such that

$$
\hat{D}^{NA}(\mathscr{X}, \mathscr{L}) \ge \delta J_Z^{NA}(\mathscr{X}, \mathscr{L})
$$

for all G_0-equivariant test configurations $(\mathscr{X}, \mathscr{L})$ for (X, L). Here a test configuration $(\mathscr{X}, \mathscr{L})$ is called G_0-*equivariant* if the natural G_0-action on $\mathscr{X}_1 = X$ extends to a G_0-action on $(\mathscr{X}, \mathscr{L})$ which covers the trivial action on $\mathbb{A}^1$ and commutes with the $\mathbb{C}^*$-action of the test configuration $(\mathscr{X}, \mathscr{L})$.

Problems

9.1 Put $N := \mathbb{P}^1(\mathbb{C}) \times \mathbb{P}^2(\mathbb{C})$ and $L := \mathscr{O}_N(1, -1)$, where for integers p and q, $\mathscr{O}_N(p, q)$ denotes the line bundle $\mathrm{pr}_1^* \, \mathscr{O}_{\mathbb{P}^1}(p) \otimes \mathrm{pr}_2^* \, \mathscr{O}_{\mathbb{P}^1}(q)$ over N with natural projections $\mathrm{pr}_i : \mathbb{P}^1(\mathbb{C}) \times \mathbb{P}^2(\mathbb{C}) \to \mathbb{P}^i(\mathbb{C})$, $i = 1, 2$. Show that $X := \mathbb{P}(\mathscr{O}_N \oplus L)$ admits no Kähler–Einstein metrics, and that X admits a generalized Kähler–Einstein metric.

9.2 For X in Problem 9.1, let $X^k := X \times \cdots \times X$ be the direct product of k copies of X. Find the smallest positive integer k such that $k \cdot \gamma_X \geq 1$. Note that, for such an integer k, the direct product X^k admits no generalized Kähler–Einstein metrics.

Appendix A
Geometry of Pseudo-Normed Graded Algebras

A.1 Differential Geometric Viewpoints

Graded algebras (such as canonical rings) coming from the space of sections of polarized algebraic varieties are studied by many mathematicians. Differential geometrically, such a study begins with a work of Royden [69], followed by the pseudo-norm project proposed by S.-T. Yau and C.-Y. Chi [16, 17]. These allow us to obtain a new aspect of the Torelli-type theorem.

In this appendix, we discuss how the geometry of L^p-spaces allows us to obtain a natural compactification of the moduli space of pseudo-normed graded algebras of a certain type. In contrast to the GIT-limits in algebraic geometry (or to the Gromov–Hausdorff limit in Riemannian geometry), we have some straightforward compactification (cf. [44]) of the moduli space of pseudo-normed graded algebras.

For stable curves in the Deligne–Mumford compactification, the notion of an orthogonal direct sum of pseudo-normed spaces comes up naturally.

A.2 L^p-Spaces

Let p be a real constant such that $p = 2$ or $0 < p \le 1$. We consider a complex vector space V of complex dimension N, where we assume $N < +\infty$ throughout subsequent sections, though only in this section, N can possibly be infinite.

Definition A.1 $(V, \|\ \|)$ is called an L^p-space, if $V \ni v \mapsto \|v\| \in \mathbb{R}_{\ge 0}$ is a continuous function, called a *pseudo-norm of order p*, satisfying the following conditions:

- For $v \in V$, we have $\|v\| = 0$ if and only if $v = 0$;
- *homogeneity*: $\|c \cdot v\| = |c| \cdot \|v\|$ for all $c \in \mathbb{C}$ and $v \in V$;

© The Author(s), under exclusive licence to Springer Nature Singapore Pte Ltd. 2021
T. Mabuchi, *Test Configurations, Stabilities and Canonical Kähler Metrics*,
SpringerBriefs in Mathematics, https://doi.org/10.1007/978-981-16-0500-0

- *subadditivity*: If $0 < p \leq 1$, then $\|u + v\|^p \leq \|u\|^p + \|v\|^p$ for all $u, v \in V$.
  If $p = 2$, then $\| \ \|$ is a Hermitian norm for V.

Hence if $p = 2$, there exists a positive definite Hermitian inner product $(,)$ such that $\|v\|^2 = (v, v)$ for all $v \in V$, and in particular $\|u + v\| \leq \|u\| + \|v\|$ for all u, $v \in V$.

Example A.1 Let $(X, d\mu)$ be a measure space. Then $V = L^p(X, d\mu)$ with the L^p-norm $\| \ \|$ is an L^p-space, where $L^p(X, d\mu)$ denotes the space of all complex measurable functions f on $(X, d\mu)$ such that $\int_X |f|^p d\mu < +\infty$.

To check the subadditivity for $0 < p \leq 1$, let $f, g \in L^p(X, d\mu)$. Recall that the inequality $(a+b)^p \leq a^p + b^p$ holds for all nonnegative real numbers a and b. Then by applying this inequality to $a = |f(x)|$ and $b = |g(x)|$, $x \in X$, we obtain the inequality $|f(x) + g(x)|^p \leq (|f(x)| + |g(x)|)^p \leq |f(x)|^p + |g(x)|^p$, and hence

$$\|f + g\|^p = \int_X |f + g|^p d\mu \leq \int_X (|f|^p + |g|^p) d\mu = \|f\|^p + \|g\|^p.$$

Example A.2 Let X be an n-dimensional compact complex connected manifold. For the canonical bundle K_X of X, we consider the dualizing sheaf $\omega_X := \mathcal{O}(K_X)$ of X. For every positive integer m, we put $p = 2/m$. Put $|\sigma|^p := (\sigma \wedge \bar{\sigma})^{1/m}$. Then $V := H^0(X, \omega_X^{\otimes m})$ has the structure of an L^p-space by the pseudo-norm

$$\|\sigma\|_m := \left(\int_X |\sigma|^p \right)^{1/p}, \qquad \sigma \in V.$$

A.3   An Orthogonal Direct Sum of L^p-Spaces

Let $(V_\lambda, \| \ \|_\lambda)$, $\lambda \in \Lambda$, be L^p-spaces for a fixed p, where $\Lambda = \{1, 2, \cdots, k\}$, and p is a real number such that $0 < p \leq 1$ or $p = 2$.

Definition A.2 $(V, \| \ \|)$ is called an *orthogonal direct sum* of L^p-spaces $(V_\lambda, \| \ \|_\lambda)$, $\lambda = 1, 2, \cdots, k$, if the following conditions are satisfied:

- $V = \bigoplus_{\lambda \in \Lambda} V_\lambda$ as a vector space.
- If $v_\lambda \in V_\lambda$, then for $v = \bigoplus_{\lambda \in \Lambda} v_\lambda$, we have

$$\|v\|^p = \sum_{\lambda \in \Lambda} \|v_\lambda\|_\lambda^p.$$

If $(V, \| \ \|)$ is an orthogonal direct sum as above, then we write

$$(V, \| \ \|) = \bigoplus_{\lambda \in \Lambda} (V_\lambda, \| \ \|_\lambda).$$

Note that an orthogonal direct sum of L^p-spaces is again an L^p-space. An L^p-space $(V, \|\ \|)$ is called *irreducible*, if it is not written as an orthogonal direct sum $(V', \|\ \|') \oplus (V'', \|\ \|'')$ of nontrivial L^p-subspaces $(V', \|\ \|')$, $(V'', \|\ \|'')$. Every L^p-space is an orthogonal direct sum of irreducible L^p-subspaces (cf. [44]). If $p = 2$, every irreducible L^p-space is 1-dimensional, and the decomposition into irreducible L^p-spaces is given by a choice of an orthonormal basis. Hence for $p = 2$, the decomposition is not unique. However, for $p \neq 2$, the decomposition is unique.

Example A.3 $(\mathbb{C}, \|\ \|)$ is an L^p-space by setting $\|z\| := |z|$ for $z \in \mathbb{C}$. We then consider its orthogonal direct sum

$$(\mathbb{C}^N, \|\ \|_N) := \bigoplus^N (\mathbb{C}, \|\ \|).$$

Hence for all $z = (z_1, \cdots, z_N) \in \mathbb{C}^N$, we have

$$\|z\|_N = (|z_1|^p + \cdots + |z_N|^p)^{1/p}.$$

For an L^p-space $(V, \|\ \|)$, the set $\Sigma := \{v \in V;\ \|v\| \leq 1\}$ is called the *indicatrix* for $(V, \|\ \|)$. In the above example, let $N = 2$. Let $0 < p \leq 1$. Then

$$\Sigma = \{(z_1, z_2) \in \mathbb{C}^2;\ |z_1|^p + |z_2|^p \leq 1\}.$$

Then Σ is, when restricted to the real points (x_1, x_2) in $\mathbb{R}^2$, described as in the following figures:

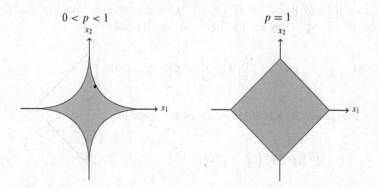

Hence for $0 < p \leq 1$, the uniqueness of the decomposition into irreducible L^p-subspaces is easily understood.

Example A.4 Let X be a compact complex simple normal crossing variety of complex dimension n. Namely, in the expression

$$X = \bigcup_{i=1}^{r} X_i,$$

irreducible components X_i, $i = 1, 2, \cdots, N$, of X are n-dimensional compact complex connected manifolds crossing normally. Let ω_X be the dualizing sheaf of X. Let $\nu : Y \to X$ be the normalization of X. For each $k \in \mathbb{Z}$ with $0 \le k \le n + 1$, we consider the disjoint union

$$Y^{[k]} := \bigsqcup_{i_0 < i_1 < \cdots < i_k} X_{i_0} \cap X_{i_1} \cap \cdots \cap X_{i_k}.$$

In particular, $Y^{[0]} = Y$ and $Y^{[n+1]} = \emptyset$. Let Res_k denote the residue map sending each $\sigma \in H^0(X, \omega_X^{\otimes m})$ to its residue $\mathrm{Res}_k(\sigma)$ along $Y^{[k]}$. Here $\mathrm{Res}_k(\sigma)$ is a meromorphic m-ple canonical form on $Y^{[k]}$ possibly with a simple pole along general points of $Y^{[k+1]}$. Then by setting

$$W_k := \mathrm{Ker}\,(\mathrm{Res}_k), \qquad k = 1, \cdots, n,$$

we have the following weight filtration of $H^0(X, \omega_X^{\otimes m})$:

$$0 \subset W_1 \subset W_2 \subset \cdots \subset W_n \subset H^0(X, \omega_X^{\otimes m}).$$

Put $W_0 := \{0\}$ and $W_{n+1} := H^0(X, \omega_X^{\otimes m})$ for convenience. We can then define $\mathrm{Gr} = \mathrm{Gr}(H^0(X, \omega_X^{\otimes m}))$ by setting $\mathrm{Gr}_k := W_{k+1}/W_k$ and

$$\mathrm{Gr} := \bigoplus_{k=0}^{n} \mathrm{Gr}_k \left(= \bigoplus_{k=0}^{n} W_{k+1}/W_k \right).$$

For $p = 2/m$, each Gr_k has the structure of L^p-space by

$$\|[\sigma]\|_k := \left(\int_{Y^{[k]}} |\mathrm{Res}_k \, \sigma|^p \right)^{1/p}, \qquad \sigma \in W_{k+1},$$

where $[\sigma] := \sigma$ modulo W_k, and we here note that $\mathrm{Res}_k \, \sigma$ belongs to $H^0(Y^{[k]}, \omega_{Y^{[k]}}^{\otimes m})$. Define $(\mathrm{Gr}, \| \; \|)$ as the orthogonal direct sum

$$(\mathrm{Gr}, \| \; \|) := \bigoplus_{k=0}^{n} (\mathrm{Gr}_k, \| \; \|_k)$$

of L^p-spaces $(\mathrm{Gr}_k, \| \; \|_k)$, $k = 0, 1, \cdots, n$.

For $n = 1$, we consider the case where X is a stable curve $C = \cup_{i=1}^{r} C_i$ of genus $g \geq 2$. Let $C_{\text{sing}} = \{x_1, \cdots, x_r\}$ be the set of the singular points in C. Let Res : $H^0(C, \omega_C^{\otimes m}) \to \mathbb{C}^r$ be the residue map defined by

$$\text{Res}(\sigma) := (\text{Res}_{x_1}(\sigma), \cdots, \text{Res}_{x_r}(\sigma)), \qquad \sigma \in H^0(C, \omega_C^{\otimes m}).$$

For $W_1 := \text{Ker Res}$ and $W_2 := H^0(C, \omega_C^{\otimes m})$, we define Gr := $\text{Gr}(H^0(C, \omega_C^{\otimes m}))$ by setting Gr $= \text{Gr}_0 \oplus \text{Gr}_1$, where $\text{Gr}_0 := W_1$ and $\text{Gr}_1 := W_2/W_1$. Put $p = 2/m$. Then

$$\|\sigma\|_0 := \left(\int_C |\sigma|^p \right)^{1/p}, \qquad \sigma \in W_1,$$

$$\|[\sigma]\|_1 := \left(\sum_{i=1}^{r} |\text{Res}_{x_i} \sigma|^p \right)^{1/p}, \qquad \sigma \in W_2,$$

where $[\sigma] := \sigma$ modulo W_1 for $\sigma \in W_2$. Recall that a Torelli-type theorem holds also for stable curves (cf. [34]).

A.4   A Boundedness Theorem for L^p-Spaces

Let $(V, \| \ \|)$ be an L^p-space of complex dimension N. For a basis $\mathbf{v} = (v_1, \cdots, v_N)$ of V, we consider the associated ellipsoid (centered at the origin)

$$E(\mathbf{v}) := \left\{ \sum_{i=1}^{N} \zeta_i v_i \ ; \ |\zeta| \leq 1, \zeta \in \mathbb{C}^N \right\},$$

where we put $|\zeta| := (\sum_{i=1}^{N} |\zeta_i|^2)^{1/2}$ for all $\zeta = (\zeta_1, \cdots, \zeta_N) \in \mathbb{C}^N$. Let $\mathscr{B}$ be the set of all bases $\mathbf{w} = (w_1, \cdots, w_N)$ for V such that $E(\mathbf{w})$ contains the indicatrix $\Sigma := \{v \in V \ ; \ \|v\| \leq 1\}$. The unitary group $U(N)$ acts on $\mathscr{B}$ from the right by

$$\mathbf{w} \cdot g := \left(\sum_{i=1}^{N} w_i g_{i1}, \sum_{i=1}^{N} w_i g_{i2}, \cdots, \sum_{i=1}^{N} w_i g_{iN} \right),$$

where $\mathbf{w} = (w_1, \cdots, w_N) \in \mathscr{B}$ and $g = (g_{ij}) \in U(N)$. Then by Demailly et al. [85, p. 663], there exists a basis $\mathbf{v}$ for V, unique up to the action of $U(N)$, such that $E(\mathbf{v})$ is the ellipsoid, containing the indicatrix Σ, of minimal volume, where the minimality is characterized by

$$\left| \frac{v_1 \wedge v_2 \wedge \cdots \wedge v_N}{w_1 \wedge w_2 \wedge \cdots \wedge w_N} \right| \leq 1, \qquad \text{for all } \mathbf{w} = (w_1, w_2, \cdots, w_N) \in \mathscr{B}.$$

$$(A.1)$$

Such a basis $\mathbf{v}$ just as above is called a *special basis* for $(V, \| \ \|)$. Consider the boundary $\partial\Sigma = \{v \in V \ ; \ \|v\| = 1\}$ of the indicatrix Σ. Let $\partial\Sigma^N := \partial\Sigma \times \cdots \times \partial\Sigma$ be the direct product of N copies of $\partial\Sigma$. Put $S^{2N-1} := \{\zeta \in \mathbb{C}^n \ ; \ |\zeta| = 1\}$. We then define

$$\mu := \inf_{\mathbf{e} \in \partial\Sigma^N} \max_{w \in \partial\Sigma} \left\{ |\zeta| \ ; \ w = \sum_{i=1}^{N} \zeta_i e_i \right\},$$

where $\mathbf{e} = (e_1, \cdots, e_N)$ is a basis for V running through the set $\partial\Sigma^N$. Then μ^{-1} in $\mathbb{R}_{\geq 0} \cup \{+\infty\}$ is just $\sup_{\mathbf{e} \in \partial\Sigma^N} (\min_{\zeta \in S^{2N-1}} \| \sum_{i=1}^{N} \zeta_i e_i \|)$. Then μ has the following upper bound (cf. [44]):

Theorem A.1 $\mu \leq \{(4^N - 1)/3\}^{1/2}$.

Proof It suffices to show the existence of a basis $\mathbf{e} = (e_1, \cdots, e_N)$ for V such that

(1) $e_i \in \partial\Sigma$ for all $i = 1, \cdots, N$;
(2) $|\zeta|^2 \leq (4^N - 1)/3$ for all $w = \sum_{i=1}^{N} \zeta_i e_i$ in $\partial\Sigma$.
Fix a basis $\mathbf{w} = (w_1, \cdots, w_N)$ for V. Put $e_0 := 0 \in V$, and for $i = 1, 2, \ldots, N$, we shall define $e_i \in \partial\Sigma$ inductively from $e_0, e_1, \cdots, e_{i-1}$ as follows: We now define a Hermitian inner product $(\ , \ )$ on V by

$$(w', w'') := \sum_{i=1}^{N} \zeta_i' \bar{\zeta}_i'',$$

for all $w' = \sum_{i=1}^{N} \zeta' w_i$ and $w'' = \sum_{i=1}^{N} \zeta'' w_i$ in V. In terms of this Hermitian inner product, let V_i be the orthogonal complement of $\mathbb{C}e_0 + \mathbb{C}e_1 + \cdots + \mathbb{C}e_{i-1}$ in V (where $e_1, e_2, \cdots, e_{i-1}$ are chosen to be linearly independent) and we consider the orthogonal projection $p_i : V \to V_i$. Since $V_1 = V$, the orthogonal projection p_1 is nothing but id_V. Put $\zeta_0 := 0$. For $0 \leq i \leq N$, we claim the following:

Claim (i): If $i = 0$, by setting $e_i = 0$, and if $i \neq 0$, by choosing a suitable $e_i \in \partial\Sigma$ such that $p_i(e_i) \neq 0$, we can write every $w \in \partial\Sigma$ as

$$w = p_{i+1}(w) + \sum_{k=0}^{i} \zeta_k e_k, \qquad (A.2)$$

such that $\zeta_k \in \mathbb{C}, k = 0, 1, \ldots, i$, satisfy $|\zeta_k| \leq 2^{i-k}$.

Then Claim (i) is obviously true for $i = 0$ by $p_1 = \mathrm{id}_V$ and $\zeta_0 = 0$. On the other hand, if Claim (i) is true for $i = N$, it then follows from $p_N(e_N) \neq 0$ that $e_1, e_2, \cdots, e_N$ are linearly independent in V, so that $V_{N+1} = \{0\}$, i.e., $p_{N+1} = 0$.

Then by (A.2) applied to $i = N$, every $w \in \partial \Sigma$ is expressible as

$$w = p_{N+1}(w) + \sum_{k=0}^{N} \zeta_k e_k = \sum_{k=1}^{N} \zeta_k e_k,$$

where $|\zeta|^2 \leq \sum_{k=1}^{N} |\zeta_k|^2 \leq \sum_{k=1}^{N} (2^{N-k})^2 = (4^N - 1)/3$, as required.

Thus the proof of Theorem A.1 is reduced to showing Claim (i) by induction on i. Hence for $1 \leq i \leq N$, we shall show Claim (i) by assuming Claim $(i-1)$. Since $\partial \Sigma \not\subset \mathbb{C} e_0 + \cdots + \mathbb{C} e_{i-1} \neq V$, and since $\partial \Sigma$ is compact, $p_i(\partial \Sigma) \neq \{0\}$ is a compact subset of V_i. Hence we can choose $e_i \in \partial \Sigma$ such that

$$(p_i(e_i), p_i(e_i)) = \max_{w \in \partial \Sigma} (p_i(w), p_i(w)) > 0. \tag{A.3}$$

In particular, $p_i(e_i) \neq 0$. By Claim $(i-1)$, every $w \in \partial \Sigma$ is expressible as

$$w = p_i(w) + \sum_{k=0}^{i-1} \xi_k e_k, \tag{A.4}$$

where $\xi_k \in \mathbb{C}$ are such that $\xi_0 = 0$ and $|\xi_k| \leq 2^{(i-1)-k}$, $1 \leq k \leq i-1$. Since $e_i \in \partial \Sigma$, applying (A.4) to $w = e_i$, we obtain

$$e_i = p_i(e_i) + \sum_{k=0}^{i-1} \eta_k e_k, \tag{A.5}$$

where $\eta_k \in \mathbb{C}$ are such that $\eta_0 = 0$ and $|\eta_k| \leq 2^{(i-1)-k}$, $1 \leq k \leq i-1$. On the other hand, by the theorem of three perpendiculars, $p_i(w)$ is written as an orthogonal sum

$$p_i(w) = p_{i+1}(w) + \alpha p_i(e_i) \tag{A.6}$$

for some $\alpha \in \mathbb{C}$. In particular, $(p_i(w), p_i(w)) \geq |\alpha|^2 (p_i(e_i), p_i(e_i))$. Now by (A.3), we see from the maximality that

$$(p_i(e_i), p_i(e_i)) \geq (p_i(w), p_i(w)) \geq |\alpha|^2 (p_i(e_i), p_i(e_i)).$$

Hence $|\alpha| \leq 1$. Then by (A.4) and (A.6),

$$w = p_{i+1}(w) + \alpha p_i(e_i) + \sum_{k=0}^{i-1} \xi_k e_k.$$

From this together with (A.5), we obtain

$$w = p_{i+1}(w) + \alpha \left(e_i - \sum_{k=0}^{i-1} \eta_k e_k \right) + \sum_{k=0}^{i-1} \xi_k e_k$$

$$= p_{i+1}(w) + \sum_{k=0}^{i} \zeta_k e_k,$$

where $\zeta_0 = \xi_0 = 0$ and for $1 \le k \le i$, $\zeta_k \in \mathbb{C}$ are defined as follows:

$$\zeta_k = \begin{cases} \xi_k - \alpha\eta_k & k = 1, 2, \cdots, i-1, \\ \alpha & k = i. \end{cases}$$

Then for $k = 0$, we have $|\zeta_k| = 0 \le 2^i$, while for $1 \le k \le i-1$,

$$|\zeta_k| = |\xi_k - \alpha\eta_k| \le |\xi_k| + |\alpha||\eta_k| \le |\xi_k| + |\eta_k| \le 2^{(i-1)-k} + 2^{(i-1)-k} = 2^{i-k}$$

and finally for $k = i$, we have $|\zeta_k| = |\alpha| \le 1 = 2^{i-k}$, as required.   □

Lemma A.1 *Let $(V, \| \ \|)$ be an L^p-space, where $0 < p \le 1$ or $p = 2$. Then there exists a real constant $C(p) \ge 1$ depending only on p such that*

$$\|u + v\| \le C(p)(\|u\| + \|v\|).$$

Proof For $p = 2$, we can choose $C(p) = 1$. For $0 < p \le 1$, we first observe that the following inequality holds for all nonnegative real numbers a, b:

$$(a^p + b^p)^{1/p} \le C(p)(a + b), \tag{A.7}$$

where $C(p) := 2^{(1/p)-1}$. In view of the subadditivity of the pseudo-norm, by applying the inequality (A.7) to $a = \|u\|$ and $b = \|v\|$, we obtain

$$\|u + v\| \le (\|u\|^p + \|v\|^p)^{1/p} \le C(p)(\|u\| + \|v\|).$$ □

For a positive integer k, let $\ell := \lceil \log_2 k \rceil$ be the round-up of $\log_2 k$, i.e., ℓ is the smallest integer such that $\log_2 k \le \ell$.

Corollary A.1 *Let $u_i \in V$, $i = 1, 2, \cdots, k$, in Lemma A.1 above. Then*

$$\left\| \sum_{i=1}^{k} u_i \right\| \le C(p)^\ell \sum_{i=1}^{k} \|u_i\|.$$

Proof For $k < i \leq 2^\ell$, we put $u_i := 0$. By applying Lemma A.1, we obtain

$$
\left\| \sum_{1 \leq i \leq 2^\ell} u_i \right\| \leq C(p) \left(\left\| \sum_{1 \leq i \leq 2^{\ell-1}} u_i \right\| + \left\| \sum_{2^{\ell-1} < i \leq 2^\ell} u_i \right\| \right) \leq \cdots \leq C(p)^\ell \sum_{i=1}^{2^\ell} \| u_i \|.
$$

Hence $\| \sum_{i=1}^k u_i \| \leq C(p)^\ell \sum_{i=1}^k \| u_i \|$, as required.   □

As to the boundedness of the moduli space of L^p spaces of a fixed dimension for a fixed p, the following theorem is fundamental:

Theorem A.2 *Let* $\mathbf{v} = (v_1, \cdots, v_N) \in \mathscr{B}$ *be a special basis for an* L^p-*space* $(V, \| \ \|)$ *of complex dimension* N, *where* $0 < p \leq 1$ *or* $p = 2$. *Then*

$$
1 \leq \left\| \sum_{i=1}^N \zeta_i v_i \right\| \leq C(N, p), \qquad \zeta = (\zeta_1, \cdots, \zeta_N) \in S^{2N-1},
$$

where $C(N, p)$ *is a positive constant depending only on* (N, p) *and is independent of the choice of* $\mathbf{v}$ *and* ζ.

Proof Since $\Sigma \subset E(\mathbf{v})$, we have $1 \leq \| \sum_{i=1}^N \zeta_i v_i \|$ for all $\zeta \in S^{2N-1}$. For an upper bound, we put $\ell := \lceil \log_2 N \rceil$. Then in view of Theorem A.1, it suffices to show

$$
\left\| \sum_{j=1}^N \zeta_j v_j \right\| \leq C(p)^{\ell(N+1)} \cdot N^{N+1} \cdot \mu^N, \qquad \zeta \in S^{2N-1}.
$$

By the definition of μ, for every $\varepsilon > 0$, we can choose a basis $\mathbf{e} = (e_1, \cdots, e_N) \in \partial \Sigma^N$ for V such that every $w = \sum_{i=1}^N \zeta_i e_i$ in $\partial \Sigma$ satisfies $|\zeta| \leq \mu + \varepsilon$, i.e., $\tilde{\mathbf{e}} := (\tilde{e}_1, \cdots, \tilde{e}_N)$ belongs to $\mathscr{B}$, where $\tilde{e}_i := (\mu + \varepsilon)e_i$. For the basis $\mathbf{e}$ for V, we can write

$$
\mathbf{v} = \mathbf{e} \cdot g
$$

for some $g \in \mathrm{GL}(N, \mathbb{C})$. Since $g = k\delta k'$ for some $k, k' \in U(N)$ and a positive real diagonal matrix δ of order N, replacing $\mathbf{v}$ by $\mathbf{v} \cdot k'^{-1}$, we may assume that $k' = 1$, i.e., $g = k\delta$. Hence

$$
v_j = \sum_{i=1}^N e_i k_{ij} \lambda_j, \qquad j = 1, 2, \ldots, N,
$$

where $k = (k_{ij}) \in U(N)$ and $\lambda_j \in \mathbb{R}_+$. Since $\Sigma \subset E(\mathbf{v})$, we have $\|v_j\| \geq 1$. Note also that $\|e_i\| = 1$ by $e_i \in \partial \Sigma$. Hence

$$1 \leq \|v_j\| = \left\| \sum_{i=1}^{N} e_i k_{ij} \lambda_j \right\| = \lambda_j \left\| \sum_{i=1}^{N} e_i k_{ij} \right\|$$

$$\leq \lambda_j C(p)^\ell \sum_{i=1}^{N} \|e_i k_{ij}\| \leq \lambda_j C(p)^\ell \sum_{i=1}^{N} \|e_i\| = \lambda_j C(p)^\ell N,$$

i.e., $\lambda_j^{-1} \leq C(p)^\ell N$ for all j. On the other hand, in view of $\mathbf{v} = \mathbf{e} \cdot (k\delta)$, it follows from $\tilde{\mathbf{e}} \in \mathscr{B}$ and (A.1) that

$$1 \geq \left| \frac{v_1 \wedge \cdots \wedge v_N}{\tilde{e}_1 \wedge \cdots \wedge \tilde{e}_N} \right| = \frac{1}{(\mu + \varepsilon)^N} \cdot \left| \frac{v_1 \wedge \cdots \wedge v_N}{e_1 \wedge \cdots \wedge e_N} \right|$$

$$= \frac{|\det k| \cdot |\det \delta|}{(\mu + \varepsilon)^N} = \frac{\prod_{i=1}^{N} \lambda_i}{(\mu + \varepsilon)^N}.$$

Hence $\prod_{i=1}^{N} \lambda_i \leq (\mu + \varepsilon)^N$. By this together with $\|v_j\| \leq \lambda_j C(p)^\ell N$ above,

$$\|v_j\| \leq \lambda_j C(p)^\ell N \leq \frac{\prod_{i=1}^{N} \lambda_i}{\prod_{i \neq j} \lambda_i} \cdot C(p)^\ell N$$

$$\leq \frac{(\mu + \varepsilon)^N}{\prod_{i \neq j} \lambda_i} \cdot C(p)^\ell N \leq (\mu + \varepsilon)^N C(p)^{\ell N} N^N.$$

Then for all $\zeta \in S^{2N-1}$, we obtain

$$\left\| \sum_{j=1}^{N} \zeta_j v_j \right\| \leq C(p)^\ell \sum_{j=1}^{N} \|\zeta_j v_j\| \leq C(p)^\ell \sum_{j=1}^{N} \|v_j\|$$

$$\leq C(p)^{\ell(N+1)} \cdot N^{N+1} \cdot (\mu + \varepsilon)^N.$$

Since $\varepsilon > 0$ is arbitrary, let $\varepsilon \to 0$. We finally obtain

$$\left\| \sum_{j=1}^{N} \zeta_j v_j \right\| \leq C(p)^{\ell(N+1)} \cdot N^{N+1} \cdot \mu^N,$$

as required.   □

A.5   The Moduli Space of L^p-Spaces

In this section, for a fixed pair (N, p) of a positive integer N and a real number p such that $0 < p \leq 1$ or $p = 2$, we shall show that the moduli space $\mathcal{M}_{N,p}$ of all linear isometric classes of L^p-spaces of complex dimension N is compact.

For $p = 2$, by a suitable choice of an orthonormal basis, it is easily seen that every $(V, \| \ \|)$ in $\mathcal{M}_{N,p}$ is linearly isometric to $(\mathbb{C}^N, \| \ \|)$ as in Sect. A.3. Hence $\mathcal{M}_{N,p}$ consists of a single point.

Let $0 < p \leq 1$, and the linearly isometric class of an L^p-space $(V, \| \ \|)$ is denoted also by the same notation $(V, \| \ \|)$ by abuse of terminology. Let

$$\beta = (V, \| \ \|) \in \mathcal{M}_{N,p} \quad \text{and} \quad \beta' = (V', \| \ \|') \in \mathcal{M}_{N,p},$$

where we choose special bases $\mathbf{v}$ and $\mathbf{v}'$ for β and β', respectively. Consider the $U(N)$-orbits $\mathbf{v} \cdot U(N)$ and $\mathbf{v}' \cdot U(N)$. Let

$$\mathbf{w} = (w_1, \cdots, w_N) \in \mathbf{v} \cdot U(N) \quad \text{and} \quad \mathbf{w}' = (w_1', \cdots, w_N') \in \mathbf{v}' \cdot U(N),$$

which are again special bases for β and β', respectively. We then consider positive real-valued functions $f_\mathbf{w}$ and $f_{\mathbf{w}'}$ on S^{2N-1} defined by

$$f_\mathbf{w}(\zeta) := \left\| \sum_{i=1}^{N} \zeta_i w_i \right\|, \qquad \zeta \in S^{2N-1},$$

$$f_{\mathbf{w}'}(\zeta) := \left\| \sum_{i=1}^{N} \zeta_i w_i' \right\|', \qquad \zeta \in S^{2N-1}.$$

Define $d(\beta, \beta') \in \mathbb{R}_{\geq 0}$ for β and β' in $\mathcal{M}_{N,p}$ by setting

$$d(\beta, \beta') := \min_{\mathbf{w}, \mathbf{w}'} \{ \max_{\zeta \in S^{2N-1}} |f_\mathbf{w}(\zeta) - f_{\mathbf{w}'}(\zeta)| \},$$

where $\mathbf{w}$ and $\mathbf{w}'$ run through the sets $\mathbf{v} \cdot U(N)$ and $\mathbf{v}' \cdot U(N)$, respectively. It is easy to check that $d(\ , \ )$ defines a distance on $\mathcal{M}_{N,p}$.

Theorem A.3 $\mathcal{M}_{N,p}$ *is compact.*

Proof Let $\beta_m = (V_m, \| \ \|_m)$, $m = 1, 2, \cdots$, be a sequence in $\mathcal{M}_{N,p}$. For each m, we can identify $V_m = \mathbb{C}^N$ by choosing a special basis $\mathbf{v}(m)$ for $(V_m, \| \ \|_m)$. Let $\mathbf{e} = (e_1, \cdots, e_N)$ be the standard basis for $\mathbb{C}^N$, i.e., $e_i := (\delta_{i1}, \delta_{i2}, \cdots, \delta_{iN}) \in \mathbb{C}^N$, where δ_{ij} is Kronecker's delta. Then by identifying $\mathbf{v}(m)$ with $\mathbf{e}$, we can write

$$\mathbf{v}(m) = (e_1, \ldots, e_N), \qquad m = 1, 2, \cdots,$$

and we put $V_\infty := \mathbb{C}^N$ and $\mathbf{v}(\infty) := (e_1, \cdots, e_N)$ for convenience. By Theorem A.2, the function $f_{\mathbf{v}(m)}(\zeta) = \|\sum_{i=1}^N \zeta_i e_i\|_m, \zeta \in S^{2N-1}$, satisfies

$$1 \leq f_{\mathbf{v}(m)}(\zeta) \leq C(N, p). \tag{A.8}$$

Let ζ and ζ' in S^{2N-1} be such that $\zeta \neq \zeta'$. Then by setting $\xi_i := (\zeta_i - \zeta_i')/|\zeta - \zeta'|$, we define $\xi := (\xi_1, \cdots, \xi_N) \in S^{2N-1}$. Then for each m,

$$|f_{\mathbf{v}(m)}(\zeta)^p - f_{\mathbf{v}(m)}(\zeta')^p| = \left| \left\| \sum_{i=1}^N \zeta_i e_i \right\|_m^p - \left\| \sum_{i=1}^N \zeta_i' e_i \right\|_m^p \right|$$

$$\leq \left\| \sum_{i=1}^N (\zeta_i - \zeta_i') e_i \right\|_m^p = \left\| \sum_{i=1}^N \xi_i e_i \right\|_m^p \cdot |\zeta - \zeta'|^p,$$

where the inequality just above follows from the subadditivity. Again by Theorem A.2, $\|\sum_{i=1}^N \xi_i e_i\|_m \leq C(N, p)$. Hence

$$|f_{\mathbf{v}(m)}(\zeta)^p - f_{\mathbf{v}(m)}(\zeta')^p| \leq C(N, p)^p \cdot |\zeta - \zeta'|^p,$$

i.e., $f_{\mathbf{v}(m)}(\zeta)^p$, $m = 1, 2, \cdots$, are Hölder continuous, of exponent p, with the Hölder norm uniformly bounded by the constant $C(N, p)^p$ independent of m. Hence by (A.8) and the compactness of S^{2N-1}, the Ascoli–Arzela theorem allows us to assume that, replacing $\{f_{\mathbf{v}(m)} ; m = 1, 2, \cdots\}$ by its subsequence if necessary, $f_{\mathbf{v}(m)}$ converges uniformly to a continuous function f_∞ ($= f_{\mathbf{v}(\infty)}(\zeta)) \geq 1$ on S^{2N-1}, as $m \to \infty$. Then for $\zeta \in \mathbb{C}^N$, we now define

$$\|\zeta\|_\infty := \begin{cases} |\zeta| f_\infty(\zeta/|\zeta|), & \zeta \neq 0, \\ 0, & \zeta = 0. \end{cases}$$

Then $\beta_\infty := (V_\infty, \| \ \|_\infty)$ is an L^p-space as follows: Note that both V_m and V_∞ are viewed as $\mathbb{C}^N$. Let $\zeta = \zeta' + \zeta''$, where $\zeta', \zeta'' \in \mathbb{C}^N$. In order to show the subadditivity, we may assume without loss of generality that $\zeta \neq 0$, $\zeta' \neq 0$, and $\zeta'' \neq 0$. For each m, by the subadditivity of $(V_m, \| \ \|_m)$, we obtain

$$\|\zeta\|_m^p \leq \|\zeta'\|_m^p + \|\zeta''\|_m^p.$$

Here the left-hand side is $(|\zeta| \cdot \|\zeta/|\zeta|\|_m)^p = |\zeta|^p f_{\mathbf{v}(m)}(\zeta/|\zeta|)^p$, while similarly the right-hand side is $|\zeta'|^p f_{\mathbf{v}(m)}(\zeta'/|\zeta'|)^p + |\zeta''|^p f_{\mathbf{v}(m)}(\zeta''/|\zeta''|)^p$. Let $m \to \infty$. Then

$$|\zeta|^p f_\infty(\zeta/|\zeta|)^p \leq |\zeta'|^p f_\infty(\zeta'/|\zeta'|)^p + |\zeta''|^p f_\infty(\zeta''/|\zeta''|)^p,$$

i.e., $\|\zeta\|_\infty^p \leq \|\zeta'\|_\infty^p + \|\zeta''\|_\infty^p$. Hence the subadditivity holds for β_∞, so that β_∞ is an L^p-space, i.e., $\beta_\infty \in \mathcal{M}_{N,p}$. To complete the proof, we shall show that $\beta_m \to$

β_∞ in $\mathcal{M}_{N,p}$, as $m \to \infty$. Since $f_\infty = f_{v(\infty)}$, in view of the inequality

$$0 \le d(\beta_m, \beta_\infty) \le \max|f_{v(m)}(\zeta) - f_{v(\infty)}(\zeta)|,$$

the right-hand side converges to 0, so that $\beta_m \to \beta_\infty$, as $m \to \infty$. $\qquad\square$

A.6   A Multiplicative System of L^p-Spaces

Let $R = \bigoplus_{m=0}^\infty R_m$ be a commutative graded $\mathbb{C}$-algebra such that (i) $R_0 = \mathbb{C}$, (ii) $\dim R_m < +\infty$ for all m, and (iii) every element r of R is written as a finite sum

$$r = \sum_m r_m$$

with $r_m \in R_m$. Moreover, there exists a multiplicative structure for R such that

$$R_\ell \times R_m \ni (r', r'') \mapsto r' \cdot r'' \in R_{\ell+m},$$

for all nonnegative integers ℓ and m. Then $(R_m, \| \ \|_m)_{m=0,1,\dots}$ is called a *multiplicative system of L^p-spaces*, if the following conditions are satisfied:

- If $r \in \mathbb{C}$, then $\|r\|_0$ is the absolute value $|r|$ of the complex number r.
- $(R_m, \| \ \|_m)$, $m = 1, 2, \cdots$, is an L^p-space for $p = 2/m$.
- *Submultiplicative inequality*: If $r' \in R_\ell$ and $r'' \in R_m$ for arbitrary positive integers ℓ and m, then $\|r' \cdot r''\|_{\ell+m} \le \|r'\|_\ell \cdot \|r''\|_m$.

Since $(R_m, \| \ \|_m)$ is an L^p-space, we observe here that, if $r \in \mathbb{C} = R_0$ and $r' \in R_m$, then $\|rr'\|_m = |r| \cdot \|r'\|_m = \|r\|_0 \cdot \|r'\|_m$.

Example A.5 For $R_m := H^0(X, \omega_X^{\otimes m})$ and $\| \ \|_m$ as in Example A.2, we consider the commutative graded algebra $R := \bigoplus_{m=0}^\infty R_m$. Then $(R_m, \| \ \|_m)_{m=0,1,\dots}$ is a multiplicative system of L^p-spaces.

More generally, let $R_m := \mathrm{Gr}(H^0(X, \omega_X^{\otimes m}))$ be as in Example A.4, and the pseudo-norm $\| \ \|$ in the same example will be denoted by $\| \ \|_m$. Then in this generalized case also, $(R_m, \| \ \|_m)_{m=0,1,\dots}$ is a multiplicative system of L^p-spaces.

Example A.6 Let $L \to X$ be a holomorphic line bundle over a compact complex connected manifold X with a volume form Ω. Let h be a Hermitian metric for L. Then $R_m := H^0(X, L^{\otimes m})$ has a structure of an L^p-space by setting

$$\|v\|_m := \left\{ \int_X |v|_h^p \, \Omega \right\}^{1/p},$$

where $p := 2/m$, and $|v|_h := \{(v, v)_h\}^{1/2}$. Then $(R_m, \| \ \|_m)_{m=0,1,\cdots}$ is a multiplicative system of L^p-spaces.

In both of these examples, as observed by Sakai (see also Yau [89]), the submultiplicative inequality is an easy consequence of the Hölder inequality as follows:
Put $p(\ell) = 2/\ell$, $p(m) := 2/m$, and $p(\ell + m) := 2/(\ell + m)$. Let $r' \in R_\ell$ and $r'' \in R_m$. Put $r := r'r'' \in R_{\ell+m}$. Then in both Examples A.5 and A.6, $\|r'\|_\ell$, $\|r''\|_m$, $\|r\|_{\ell+m}$ are viewed as $\|r'\|_{L^{p(\ell)}}$, $\|r''\|_{L^{p(m)}}$, $\|r\|_{L^{p(\ell+m)}}$, respectively. Since

$$
\frac{1}{p(\ell)} + \frac{1}{p(m)} = \frac{\ell}{2} + \frac{m}{2} = \frac{1}{p(\ell + m)},
$$

the Hölder inequality shows that $\|r' \cdot r''\|_{L^{p(\ell+m)}} \leq \|r'\|_{L^{p(\ell)}} \|r''\|_{L^{p(m)}}$. Hence

$$
\|r' \cdot r''\|_{\ell+m} = \|r' \cdot r''\|_{L^{p(\ell+m)}} \leq \|r'\|_{L^{p(\ell)}} \|r''\|_{L^{p(m)}} = \|r'\|_\ell \cdot \|r''\|_m.
$$

To each $m = 1, 2, \cdots$, we assign a nonnegative integer N_m. Let $\rho^{(j)}$, $j = 1, 2, \cdots$, be a sequence of multiplicative systems of L^p-spaces such that

- $\rho^{(j)} := (R_m^{(j)}, \| \ \|_m^{(j)})_{m=0,1,\cdots}$ for all j;
- $\dim R_m^{(j)} = N_m$ for all m and j.

Since $\mathscr{M}_{N_1,p}$ consists of a single point for $p = 2$ (i.e., $m = 1$) by choosing an orthonormal basis for each $R_1^{(j)}$, we have the identification

$$
(R_1^{(1)}, \| \ \|_1^{(1)}) = (R_1^{(2)}, \| \ \|_1^{(2)}) = \cdots = (R_1^{(j)}, \| \ \|_1^{(j)}) = \cdots
$$

in $\mathscr{M}_{N_1,2}$. Hence let $m \geq 2$. By $p = 2/m$, we have $0 < p \leq 1$. Then by the compactness of the moduli spaces of L^p-spaces (see Theorem A.3), replacing $\{j = 1, 2, \cdots\}$ by its subsequence $\{j_1, j_2, \cdots, j_k, \cdots\}$ if necessary, we may assume that

$$
(R_2^{(j_k)}, \| \ \|_2^{(j_k)}) \to (R_2^{(\infty)}, \| \ \|_2^{(\infty)}) \text{ in } \mathscr{M}_{N_2,1}, \quad \text{as } k \to \infty,
$$

for some $(R_2^{(\infty)}, \| \ \|_2^{(\infty)})$ in $\mathscr{M}_{N_2,1}$. Next replacing $\{j_1, j_2, \cdots, j_k, \cdots\}$ by its subsequence $\{j_{k_1}, j_{k_2}, \cdots, j_{k_\alpha}, \cdots\}$ if necessary, we may assume that

$$
(R_3^{(j_{k_\alpha})}, \| \ \|_3^{(j_{k_\alpha})}) \to (R_3^{(\infty)}, \| \ \|_3^{(\infty)}) \text{ in } \mathscr{M}_{N_3,2/3}, \quad \text{as } \alpha \to \infty,
$$

for some $(R_3^{(\infty)}, \| \ \|_3^{(\infty)})$ in $\mathscr{M}_{N_3, 2/3}$. We repeat this process. Hence, starting from $\{1, 2, 3, \cdots\}$, we have successive subsequences as follows:

$$\{1, \quad 2, \quad 3, \cdots, \ j, \quad \cdots\},$$
$$\{j_1, \ j_2, \ j_3, \cdots, \quad j_k, \quad \cdots\},$$
$$\{j_{k_1}, \ j_{k_2}, \ j_{k_3}, \cdots, \ j_{k_\alpha}, \cdots\},$$

and the diagonal sequence $1, \ j_2, \ j_{k_3}, \ \cdots$ will be rewritten as $i_1, i_2, i_3, \cdots$, so that for $p(m) = 2/m$, by replacing the original sequence $\{1, 2, \cdots\}$ by its subsequence $\{i_1, i_2, \cdots\}$, we may assume from the beginning that, for each $m = 1, 2, \cdots$,

$$(R_m^{(j)}, \| \ \|_m^{(j)}) \to (R_m^{(\infty)}, \| \ \|_m^{(\infty)}) \text{ in } \mathscr{M}_{N_m, p(m)}, \quad \text{as } j \to \infty.$$

The multiplicative structure for $R^{(\infty)} := \bigoplus_{m=1}^\infty R_m^{(\infty)}$ is defined as follows: For each j, we consider the multiplicative system

$$(R_m^{(j)}, \| \ \|_m^{(j)})_{m=0,1,\cdots}$$

of L^p-spaces, and we choose a special basis $\mathbf{v}(m; j) = (v(m; j)_1, \cdots, v(m; j)_{N_m})$ for $(R_m^{(j)}, \| \ \|_m^{(j)})$. Hence for each integer $\ell > 0, \mathbf{v}(\ell; j) = (v(\ell; j)_1, \cdots, v(\ell; j)_{N_\ell})$ denotes a special basis for $(R_\ell^{(j)}, \| \ \|_\ell^{(j)})$, while $\mathbf{v}(\ell + m; j) = (v(\ell + m; j)_1, \cdots, v(\ell + m; j)_{N_{\ell+m}})$ denotes a special basis for $(R_{\ell+m}^{(j)}, \| \ \|_{\ell+m}^{(j)})$. In terms of these bases, the multiplicative structure of $R^{(j)}$ is given by

$$v(\ell; j)_\alpha \cdot v(m; j)_\beta = \sum_{\gamma=1}^{N_{\ell+m}} C_{\alpha,\beta}^\gamma(j) \, v(\ell + m; j)_\gamma, \tag{A.9}$$

where $C_{\alpha,\beta}^\gamma(j)$ are called the *structure constants* for the special bases. Given a multiplicative structure, these constants depend only on the choice of the special bases, so that they are unique up to the action of $U(N_\ell) \times U(N_m) \times U(N_{\ell+m})$. By Theorem A.2, we have the inequalities

$$1 \leq \| v(\ell; j)_\alpha \|_\ell^{(j)} \leq C(N_\ell, \ p(\ell)), \tag{A.10}$$

$$1 \leq \| v(m; j)_\beta \|_m^{(j)} \leq C(N_m, \ p(m)). \tag{A.11}$$

Put $|C_{\alpha,\beta}(j)| := (\sum_{\gamma=1}^{N_{\ell+m}} |C_{\alpha,\beta}^\gamma(j)|^2)^{1/2}$ and $\xi_\gamma := C_{\alpha,\beta}^\gamma(j)/|C_{\alpha,\beta}(j)|$. Then by the submultiplicative inequality together with (A.9), (A.10) and (A.11), we obtain

$$C(N_\ell, p(\ell)) \cdot C(N_m, p(m)) \geq \| v(\ell; j)_\alpha \|_\ell^{(j)} \cdot \| v(m; j)_\beta \|_m^{(j)}$$

$$\geq \| v(\ell; j)_\alpha \cdot v(m; j)_\beta \|_{\ell+m}^{(j)} = \left\| \sum_{\gamma=1}^{N_{\ell+m}} C_{\alpha,\beta}^\gamma(j)\, v(\ell+m; j)_\gamma \right\|_{\ell+m}^{(j)}$$

$$= |C_{\alpha,\beta}(j)| \cdot \left\| \sum_{\gamma=1}^{N_{\ell+m}} \xi_\gamma\, v(\ell+m; j)_\gamma \right\|_{\ell+m}^{(j)} \geq |C_{\alpha,\beta}(j)|,$$

where the last inequality follows from Theorem A.2. This uniform boundedness for $|C_{\alpha,\beta}(j)|$ shows that, replacing $\{j = 1, 2, \cdots\}$ by its suitable subsequence if necessary, we may assume that, for each pair (ℓ, m) of positive integers, there exists a limit $C_{\alpha,\beta}^\gamma(\infty)$ in the sense that

$$C_{\alpha,\beta}^\gamma(j) \to C_{\alpha,\beta}^\gamma(\infty), \quad \text{as } j \to \infty.$$

Hence, the multiplicative structure of $R^{(\infty)} = \bigoplus_{m=0}^\infty R_m^{(\infty)}$ is defined by the structure constants $\{C_{\alpha,\beta}^\gamma(\infty)\}$ thus obtained. This shows the compactness of the moduli space of all multiplicative systems of L^p-spaces.

A.7   Degeneration Phenomena

Consider a proper holomorphic map of complex manifolds

$$f : \mathscr{X} \to \Delta := \{t \in \mathbb{C};\ |t| < \varepsilon\}$$

as in Masur [57] satisfying the following conditions:

1. The central fiber $\mathscr{X}_0$ is a stable curve of genus g.
2. $\mathscr{X}_t, t \neq 0$, is a nonsingular irreducible projective algebraic curve of genus $g \geq 2$.

Let $p(m) = 2/m$. Then for each $t \neq 0$, let $\| \ \|_{\mathscr{X}_t, m}$ be the pseudo-norm for $H^0(\mathscr{X}_t, \omega_{\mathscr{X}_t}^{\otimes m})$ as in Example A.2, while for $t = 0$, we consider the pseudo-norm $\| \ \|_{\mathscr{X}_0, m}$ for $\mathrm{Gr}(H^0(\mathscr{X}_0, \omega_{\mathscr{X}_0}^{\otimes m}))$ as in Example A.4. Put $N_m := \dim H^0(\mathscr{X}_t, \omega_{\mathscr{X}_t}^{\otimes m})$. Then

$$(H^0(\mathscr{X}_t, \omega_{\mathscr{X}_t}^{\otimes m}), \| \ \|_{\mathscr{X}_t, m}) \to (\mathrm{Gr}(H^0(\mathscr{X}_0, \omega_{\mathscr{X}_0}^{\otimes m})), \| \ \|_{\mathscr{X}_0, m}), \quad \text{as } t \to 0,$$

in $\mathcal{M}_{N_m, p(m)}$. Moreover, for the associated multiplicative systems of L^p-spaces, we have the following convergence:

$$\bigoplus_{m=1}^{\infty} H^0(\mathcal{X}_t, \omega_{\mathcal{X}_t}^{\otimes m}) \;\to\; \bigoplus_{m=1}^{\infty} \mathrm{Gr}(H^0(\mathcal{X}_0, \omega_{\mathcal{X}_0}^{\otimes m})),$$

as $t \to 0$. It is easily seen that a higher-dimensional analogue of such a convergence for degeneration is also true.

Solutions

Problems of Chap. 1

1.1 Let $\ddot{f}(0) = 0$. Then by the second variation formula (1.9), the vector field along X associated to the fundamental generator of σ is tangent to X in $\mathbb{P}(V^*)$. Hence the one-parameter group $\sigma : \mathbb{C}^* \to \mathrm{SL}(V)$ preserves the subvariety X in $\mathbb{P}(V^*)$.

1.2 Since ω is in the class $c_1(X)$, $\mathrm{Ric}(\omega) = \omega + dd^c f$ for some real-valued smooth function f. Then by the notation in Sect. 1.1, the scalar curvature is written as

$$S_\omega = \sum_{\alpha,\beta} g^{\bar{\beta}\alpha} R_{\alpha\bar{\beta}} = n + \Delta_\omega f,$$

where $\Delta_\omega = \sum_{\alpha,\beta} g^{\bar{\beta}\alpha} \partial^2/\partial z_\alpha \partial z_{\bar{\beta}}$. On the other hand, since ω is CSC Kähler, we have $S_\omega = n$. Hence $\Delta_\omega f = 0$. Note that every harmonic function on a smooth irreducible projective variety is constant. Thus f is constant, and $\mathrm{Ric}(\omega) = \omega$.

Problems of Chap. 2

2.1 In the definition of the ideal I, by setting $z = 1$, we have the ideal

$$I_1 = ((x_0 + x_3)x_3 - x_2^2, \; x_0(x_0 + x_3) - x_1 x_2, \; x_0 x_2 - x_1 x_3, \; x_1^2 x_3 - x_0^2(x_0 + x_3))$$

© The Author(s), under exclusive licence to Springer Nature Singapore Pte Ltd. 2021
T. Mabuchi, *Test Configurations, Stabilities and Canonical Kähler Metrics*,
SpringerBriefs in Mathematics, https://doi.org/10.1007/978-981-16-0500-0

for $\mathscr{X}_1$. Let $p \in \mathscr{X}_1$. If $x_1(p) = x_3(p) = 0$, then $x_2(p)^2 = x_0(p)^2 = 0$ in contradiction. Hence by setting $U := \{ p \in \mathscr{X}_1 ; x_1(p) \neq 0 \}$ and $V := \{ p \in \mathscr{X}_1 ; x_3(p) \neq 0 \}$, we obtain $U \cup V = \mathscr{X}_1$. Put $\xi := x_0/x_1$ and $\eta := x_2/x_3$. By $x_0 x_2 - x_1 x_3 = 0$,

$$\xi \eta = 1.$$

Then $x_2/x_1 = \xi^2/(1 - \xi^2)$ and $x_3/x_1 = \xi^3/(1 - \xi^2)$ on U, where $\xi = \pm 1$ correspond to the points $(0 : 0 : 1 : \pm 1)$, respectively. On the other hand, on V, we have $x_0/x_3 = \eta^2 - 1$ and $x_1/x_3 = (\eta^2 - 1)\eta$. Thus $U = \{\xi \in \mathbb{C}\} \cong \mathbb{C}$ and $V = \{\eta \in \mathbb{C}\} \cong \mathbb{C}$, and $\mathscr{X}_1 = U \cup V$ is isomorphic to $\mathbb{P}^1(\mathbb{C})$.

2.2 Let an integer m be such that $m \gg 1$. Note that $N_m := \dim H^0(X, L^{\otimes m})$ is $3m + 1$. In view of $H^0(\mathscr{X}_0, \mathscr{L}_0) = \Sigma_{i=0}^3 \mathbb{C} x_i$, since the map $H^0(\mathscr{X}_0, \mathscr{L}_0)^{\otimes m} \to H^0(\mathscr{X}_0, \mathscr{L}_0^{\otimes m})$ is surjective, it follows from (2.14) that

$$H^0(\mathscr{X}_0, \mathscr{L}_0^{\otimes m}) = \mathbb{C} x_2 x_3^{m-1} \oplus V,$$

where V is a $3m$-dimensional space generated by x_0, x_1 and x_3. Since $\mathbb{C}^*$ acts on V trivially, the weight w_m of the $\mathbb{C}^*$-action on $\det H^0(\mathscr{X}_0, \mathscr{L}_0^{\otimes m})$ is just the weight of the $\mathbb{C}^*$-action on $x_2 x_3^{m-1}$ viewed as a section as above, and hence $w_m = -1$. Since

$$\frac{w_m}{m N_m} = \frac{-1}{m(3m + 1)} = \frac{-1}{3m^2} \sum_{i=0}^{\infty} \left(\frac{-1}{3m} \right)^i,$$

it then follows that $\mathrm{DF}_1(\mathscr{X}, \mathscr{L}) = 0$.

2.3 Let $Y := \{(x_0 : x_1 : x_3) \in \mathbb{P}^2(\mathbb{C}) ; x_1^2 x_3 = x_0^2(x_0 + x_3)\}$ be the subvariety of $\mathbb{P}^2(\mathbb{C})$ on which $\mathbb{C}^*$ acts trivially. Then $(0 : 0 : 1)$ is the singular point of Y, and the mapping

$$\mathbb{P}^1(\mathbb{C}) \ni y := (y_0 : y_1) \mapsto \sigma(y) := ((y_1^2 - y_0^2)y_0 : (y_1^2 - y_0^2)y_1 : y_0^3) \in Y$$

is the normalization of Y such that $\sigma^{-1}((0 : 0 : 1)) = \{(1 : \pm 1)\}$. Put $F := \mathscr{O}_{\mathbb{P}^2}(1)_{|Y}$. Let $(\mathscr{Y}, \mathscr{F})$ be the trivial test configuration for (X, F) defined by

$$\mathscr{Y} := Y \times \mathbb{A}^1, \qquad \mathscr{F} := \mathrm{pr}_1^* F,$$

where $\mathrm{pr}_1 : Y \times \mathbb{A}^1 \to Y$ denotes the projection to the first factor. Then the projection $\mathbb{P}^3(\mathbb{C}) \times \mathbb{A}^1 \ni ((x_0 : x_1 : x_2 : x_3), z) \mapsto ((x_0 : x_1 : x_3), z) \in \mathbb{P}^2(\mathbb{C}) \times \mathbb{A}^1$ induces a $\mathbb{C}^*$-equivariant birational morphism $p : \mathscr{X} \to \mathscr{Y}$ such that $\mathscr{L} = p^* \mathscr{F}$. Put

$$\tilde{\mathscr{Y}} := \mathbb{P}^1(\mathbb{C}) \times \mathbb{A}^1, \quad \tilde{\sigma} := \sigma \times \mathrm{id}_{\mathbb{A}^1}, \quad \tilde{\mathscr{F}} := \tilde{\sigma}^* \mathscr{F}.$$

Obviously, $(\tilde{\mathscr{Y}}, \tilde{\mathscr{F}})$ is a trivial test configuration for $(X, L) = (\mathbb{P}^1(\mathbb{C}), \mathscr{O}_{\mathbb{P}^1}(3))$. Since $\tilde{\sigma} : \tilde{\mathscr{Y}} \to \mathscr{Y}$ lifts to a normalization $\nu : \tilde{\mathscr{Y}} \to \mathscr{X}$ such that $p \circ \nu = \tilde{\sigma}$ and that $(\tilde{\mathscr{X}}, \tilde{\mathscr{L}}) = (\tilde{\mathscr{Y}}, \tilde{\mathscr{F}})$. We now conclude that the test configuration $(\mathscr{X}, \mathscr{L})$ is trivial.

Problems of Chap. 3

3.1 For a K-invariant real symplectic form ω' on X cohomologous to ω, we consider the real-valued smooth function f' on X such that

$$df' = i_y \omega'.$$

Since ω and ω' are cohomologous, there exists a real 1-form η such that $\omega' - \omega = d\eta$. Since both ω and ω' are K-invariant, we obtain

$$\omega' - \omega = \frac{1}{|K|} \int_K (k^*\omega')\, dk - \frac{1}{|K|} \int_K (k^*\omega)\, dk = \frac{1}{|K|} \int_K (k^*d\eta)\, dk$$
$$= \frac{1}{|K|} \int_K d(k^*\eta)\, dk = d\left\{ \frac{1}{|K|} \int_K (k^*\eta)\, dk \right\},$$

where dk is the invariant measure for K, and $|K|$ is the total measure for K. Since $(1/|K|) \int_K (k^*\eta) dk$ is a K-invariant 1-form on X, we may assume from the beginning that η is K-invariant. In particular, $L_y \eta = 0$. Then

$$d(f' - f) = i_y(\omega' - \omega) = i_y d\eta = L_y \eta - d i_y \eta = d(-i_y \eta).$$

Hence, up to an additive constant, f' coincides with $f - i_y \eta$, so that we may assume, without loss of generality, that $f' = f - i_y \eta$. By $df = i_y \omega$, the vector field y vanishes at the critical points of f, and the same thing is true also f'. Keeping this in mind, let g and g' be the restrictions of f and f', respectively, to the zero set $Z := \{ p \in X ;\ y(p) = 0 \}$ of y. Then

$$\max_X f - \min_X f = \max_Z g - \min_Z g \quad \text{and} \quad \max_X f' - \min_X f' = \max_Z g' - \min_Z g'.$$

Since $f' = f - i_y \eta$, and since $i_y \eta$ vanishes on Z, we have $g = g'$. Finally, we obtain $\max_X f - \min_X f = \max_X f' - \min_X f'$, as required.

3.2 Let $\bar{\partial}^*$ be the formal adjoint, with respect to the Kähler metric ω, for the $\bar{\partial}$-operator on the space of C^∞ differential forms on X. Note that $y = \mathrm{grad}_\omega^{\mathbb{C}} \psi$. Then

$$\int_X (\Delta_{\tilde{\omega}} \xi)\, \eta\, \tilde{\omega}^n = \int_X \{\Delta_\omega \xi - \sqrt{-1}\dot{\zeta}(\psi)\, \bar{y}\, \xi\}\, \eta\, e^{\zeta(\psi)}\, \omega^n$$

$$= -\int_X (\bar{\partial}^* \bar{\partial} \xi)\, \eta\, e^{\zeta(\psi)}\, \omega^n + \int_X \dot{\zeta}(\psi) \sum_{\alpha,\beta} g^{\bar{\beta}\alpha} \frac{\partial \psi}{\partial z_\alpha} \frac{\partial \xi}{\partial z_{\bar{\beta}}}\, \eta\, e^{\zeta(\psi)}\, \omega^n$$

$$= -\int_X (\bar{\partial}\xi, \bar{\partial}\{\eta e^{\zeta(\psi)}\})_\omega\, \omega^n + \int_X (\bar{\partial}\xi, \bar{\partial}e^{\zeta(\psi)})_\omega\, \eta\, \omega^n$$

$$= -\int_X (\bar{\partial}\xi, \bar{\partial}\eta)_\omega\, e^{\zeta(\psi)}\, \omega^n = -\int_X (\bar{\partial}\xi, \bar{\partial}\eta)_\omega\, \tilde{\omega}^n.$$

3.3 Since ω is a constant scalar curvature metric, we have $\Delta_\omega f_\omega = S_\omega - S_0 = 0$. Hence for $y = \mathrm{grad}_\omega^{\mathbb{C}} \psi$, we obtain

$$\mathscr{F}(y) = \int_X \{(\mathrm{grad}_\omega^{\mathbb{C}} \psi) f_\omega\}\omega^n = \frac{1}{\sqrt{-1}} \int_X (\bar{\partial}\psi, \bar{\partial} f_\omega)_\omega \omega^n$$

$$= \frac{1}{\sqrt{-1}} \int_X \psi(\Delta_\omega f_\omega)\omega^n = 0.$$

Problems of Chap. 4

4.1 In view of the solution of Problem 2.2, $N_m = 3m + 1$, $\gamma_m = m\gamma = m$, and the weights of the $\mathbb{C}^*$-action on $H^0(\mathscr{X}_0, \mathscr{L}_0^{\otimes m})$ are $-1, 0, \ldots, 0$, i.e.,

$$\beta_1 = 1 \quad \text{and} \quad \beta_i = 0 \quad \text{for } 1 < i \le N_m.$$

Then their average is $\beta_0 := N_m^{-1} \Sigma_{i=1}^{N_m} \beta_i$ is $(3m + 1)^{-1}$. Hence $\tilde{\beta}_1 := \beta_1 - \beta_0 = 1 - (3m + 1)^{-1}$ and $\tilde{\beta}_i := \beta_i - \beta_0 = -(3m + 1)^{-1}$ for $1 < i \le N_m$. Thus

$$\|\mu\|_{asymp} = \lim_{m \to \infty} \left(\gamma_m^{-1} N_m^{-1} \sum_{i=1}^{N_m} |\tilde{\beta}_i| \right) = 0.$$

4.2 Since the test configuration $\mu = (\mathscr{X}, \mathscr{L})$ is a product configuration, $H^0(\mathscr{X}_0, \mathscr{L}_0^{\otimes m})$ is identified with $H^0(X, L^{\otimes m}) = \oplus_{i=0}^m \mathbb{C}x_0^i x_1^{m-i}$, where the weight β_i of the $\mathbb{C}^*$-action on $x_0^i x_1^{m-i}$ is $i - (m - i) = 2i - m$. Hence all the weights are

$$-m, -m+2, -m+4, \cdots, m-4, m-2, m.$$

Since their average is 0, we have $\tilde{\beta}_i := \beta_i - 0 = 2i - m$. Note also that $N_m = m + 1$ and $\gamma_m = m$. Then we can compute the asymptotic ℓ_1-norm of μ as follows:

$$\|\mu\|_{asymp} = \lim_{m \to \infty} \left(\gamma_m^{-1} N_m^{-1} \sum_{i=0}^{m} |\tilde{\beta}_i| \right) = \lim_{m \to \infty} \frac{\sum_{i=0}^{m} |2i - m|}{m(m+1)} = \frac{1}{2}.$$

4.3 For each fixed $s \in \mathbb{R}$, we have $\varliminf_{j \to \infty} \dot{f}_j(s) \leq \varliminf_{k \to \infty} \dot{f}_{j_k}(s)$ in (4.5). From this inequality, by letting $s \to -\infty$, we obtain $F_1(\{\mu_j\}) \leq F_1(\{\mu_{j_k}\})$, as required.

Problems of Chap. 5

5.1 Assume that (X, L) is strongly K-stable. For a nontrivial normal test configuration $\mu = (\mathscr{X}, \mathscr{L})$, of exponent γ, let $\{\mu_j\}$ be the sequence of test configurations as in (5.28). In view of Remark 4.5, by setting $\tilde{\mu}_j = \mu_{\tilde{\sigma}_j}$, we have a special one-parameter group $\tilde{\sigma}_j : \mathbb{C}^* \to \mathrm{SL}(V_{\gamma_j})$ such that

$$F_1(\{\mu_j\}) = F_1(\{\tilde{\mu}_j\}) \tag{5.30}$$

and that $\tilde{\sigma}_j$ is obtained as a lift of the special linearization σ_j^{SL} of σ_j to a suitable covering algebraic torus. Then by (5.29),

$$\mathrm{DF}_1(\mu) \leq \frac{\|\mu\|_{asymp}}{(n+1)c_1(L)^n[X]} F_1(\{\tilde{\mu}_j\}) \leq 0,$$

where the last inequality follows from the strong K-semistability of (X, L). Hence (X, L) is K-semistable.

Next, let $\mu := (\mathscr{X}, \mathscr{L})$ be a normal test configuration, of exponent γ, for (X, L) such that $\mathrm{DF}_1(\mu) = 0$, where for contradiction we assume that μ is nontrivial. Then by the nontriviality of μ, the associated special one-parameter group

$$\sigma : \mathbb{C}^* \to \mathrm{GL}(V_\gamma)$$

satisfying $\mu = \mu_\sigma$ is nontrivial. We observe here that the nontriviality of σ is characterized by the distinctness between the highest weight and the lowest weight for the fundamental generator of σ. Hence σ_j, $j = 1, 2, \cdots$, induced by σ are also all nontrivial. Then σ_j^{SL}, $j = 1, 2, \cdots$, are also nontrivial and so are $\tilde{\sigma}_j$. Hence

$$\{\tilde{\mu}_j\} \in \mathscr{M}.$$

Now by the strong K-stability of (X, L), we obtain $F_1(\{\tilde{\mu}_j\}) < 0$. From this together with (5.29) and (5.30), we obtain $\mathrm{DF}_1(\mu) < 0$ in contradiction, as required.

5.2 For contradiction, we assume that such an N doesn't exists. Then we have an increasing sequence of integers

$$1 \leq j_1 < j_2 < \cdots < j_\alpha < \cdots$$

such that σ_{j_α}, $\alpha = 1, 2, \cdots$, are all nontrivial. Then $\{\mu_{j_\alpha}\}_{\alpha=1,2,\cdots} \in \mathcal{M}$. Hence by the strong K-stability of (X, L), we see that $F_1(\{\mu_{j_\alpha}\}) < 0$. On the other hand, by Problem 4.3, we have $0 = F_1(\{\mu_j\}) \leq F_1(\{\mu_{j_\alpha}\})$ in contradiction, as required.

Problems of Chap. 6

6.1 By $T \subset T'$, we have $1\mathrm{PS}(T'^{\perp}_\gamma) \subset 1\mathrm{PS}(T^{\perp}_\gamma)$ for all γ. Assuming that (X, L) is uniformly K-stable relative to T, let $\sigma \in 1\mathrm{PS}(T'^{\perp}_\gamma)$. Then $\sigma \in 1\mathrm{PS}(T^{\perp}_\gamma)$. Then by the uniform K-stability of (X, L) relative to T, $\mu_\sigma := (\mathscr{X}_\sigma, \mathscr{L}_\sigma)$ satisfies

$$\mathrm{DF}_1(\mu_\sigma) \leq -C\|\mu_\sigma\|^T_{asymp},$$

where C is a positive real constant independent of γ and σ. Note that, by $T \subset T'$, we have $\|\mu_\sigma\|^{T'}_{asymp} \leq \|\mu_\sigma\|^T_{asymp}$. Hence

$$\mathrm{DF}_1(\mu_\sigma) \leq -C\|\mu_\sigma\|^{T'}_{asymp},$$

i.e., (X, L) is uniformly K-stable relative to T', as required.

6.2 Assume that Conjecture (1) is true. We first show the direction ($\Rightarrow$) in Conjecture (2), i.e., assuming further $\mathscr{E}_{\mathrm{CSC}} \neq \emptyset$, we shall show the statement on the right-hand side of (2). $\mathscr{F} = 0$ follows from $\mathscr{E}_{\mathrm{CSC}} \neq \emptyset$ by the obstruction result of Calabi and Futaki, while by (1) and $\emptyset \neq \mathscr{E}_{\mathrm{CSC}} \subset \mathscr{E}_{\mathrm{EX}}$, we also see that (X, L) is uniformly K-stable relative to T_{max}.

Next we shall show the converse direction in Conjecture (2). Then by (1), in view of the statement on the right-hand side of (2), we obtain $\mathscr{E}_{\mathrm{EX}} \neq \emptyset$ and we have an extremal Kähler metric ω in the class $c_1(L)$. Now by $\mathscr{F} = 0$, the extremal vector field of ω is zero, i.e., S_ω is constant. Hence $\mathscr{E}_{\mathrm{CSC}} \neq \emptyset$.

Problems of Chap. 7

7.1 By Sect. 7.1, since X admits a CSC Kähler metric in $c_1(L)$, the polarized algebraic manifold (X, L) is strongly K-semistable. Then by the same argument as in the solution of Problem 5.1, the strong K-semistability of (X, L) implies the K-semistability of (X, L), as required.

7.2 No. For the case $L = K_X^{-1}$, an example of Ono, Sano and Yotsutani (see [66]) is known: There exists a toric Fano Kähler–Einstein manifold X such that (X, K_X^{-1}) is not asymptotically Chow stable. Such an example was studied by Nill and Paffenholz (see [64]) from another viewpoints. In the proof of the asymptotic Chow unstability, the obstruction by Futaki (see [25]) is used.

Problems of Chap. 8

8.1 Let Λ be the maximal connected linear algebraic subgroup of $\mathrm{Aut}^0(X)$ (see Fujiki [22]). By Mumford et al. [63], the action of Λ on X lifts to a linearization of $L^{\otimes \gamma}$ for some positive integer γ. Then if $X \subset \mathbb{P}(V_\gamma^*)$ is Chow polystable, the Matsushima–Lichnerowicz obstruction vanishes as follows: (In particular, if (X, L) is asymptotically Chow polystable, then the Matsushima–Lichnerowicz obstruction vanishes.)

Let $0 \neq \hat{X} \in W_\gamma$ be the Chow form for the image cycle of X in $\mathbb{P}(V_\gamma^*)$. Put $G := \mathrm{SL}(V_\gamma^*)$. Then by the Chow polystability of $X \subset \mathbb{P}(V_\gamma^*)$, the orbit $G \cdot \hat{X}$ is closed in W_γ, and in particular affine. Hence by Matsushima's theorem [59], the isotropy subgroup H of G at $\hat{X}$ is reductive, i.e., the identity component H^0 of H is reductive. By sending $\lambda \in \Lambda$ to $g(\lambda) \in G$ induced by λ, we have an inclusion

$$g : \Lambda \hookrightarrow G, \qquad \lambda \mapsto g(\lambda).$$

Since $H^0 \subset g(\Lambda)$, we see that $\dim \Lambda - \dim H^0 = 0$ or 1. Hence either $\dim \Lambda = \dim H^0$ or the algebraic torus $g(\Lambda)/H^0 \cong \mathbb{C}^*$ induces the isotropy representation of the line $\mathbb{C}\hat{X}$ at the Chow point $[\hat{X}]$ of X. In both cases, since H is reductive, Λ is reductive, i.e., the Matsushima–Lichnerowicz obstruction vanishes.

8.2 A complete solution of the existence problem of Kähler–Einstein metrics with positive Ricci curvature for compact complex surfaces was given by Tian [78] (see also Tian and Yau [81], Siu [74]). All compact complex connected surfaces S other than $\mathbb{P}^2(\mathbb{C})$ and $\mathbb{P}^1(\mathbb{C}) \times \mathbb{P}^2(\mathbb{C})$ which has Kähler–Einstein metrics of positive Ricci curvature are the del Pezzo surfaces of degree $d := c_1(S)^2[S]$ with $1 \leq d \leq 6$. Here S is called *del Pezzo* if anticanonical bundle K_S^{-1} is ample. Topologically they are

$$\mathbb{P}^2(\mathbb{C}) \# k \, \overline{\mathbb{P}^2(\mathbb{C})}, \qquad 3 \leq k \leq 8,$$

where $k := 9 - d$. Such surfaces are obtained from the complex projective plane $\mathbb{P}^2(\mathbb{C})$ by blowing up k points in a "general position" in $\mathbb{P}^2(\mathbb{C})$. For instance, if $k = 3$, there is only one such surface which is just $\mathbb{P}^2(\mathbb{C})$ blown up at three non-collinear points.

Problems of Chap. 9

9.1 Clearly, N admits a Kähler–Einstein metric ϕ in the class $c_1(N) = c_1(\mathcal{O}_N(2,3))$. By $L = \mathcal{O}_N(1,-1)$, we can choose a Hermitian metric h for L such that the eigenvalues β_i ($i = 1, 2, 3$) of $\mathrm{Ric}(h)$ with respect to ϕ are constant. Namely, $(\beta_1, \beta_2, \beta_3) = (1/2, -1/3, -1/3)$. Then by $b_\alpha = \int_{-1}^1 u^\alpha \prod_{i=1}^3 (1 - \beta_i u) du$, $\alpha = 1, 2$,

$$(b_1, b_2) = (4/45, 26/45).$$

Since $b_1 \neq 0$, by a result of Koiso and Sakane [36], X admits no Kähler–Einstein metrics. Moreover, by $|b_1| < b_2$, we see from Theorem 9.5 that X admits a generalized Kähler–Einstein metric.

9.2 Note that X in Problem 9.1 admits a generalized Kähler–Einstein metric by $(b_1, b_2) = (4/45, 26/45)$. Moreover, $b_0 = \int_{-1}^1 \prod_{i=1}^3 (1 - \beta_i u) du = 50/27$. Then by Remark 9.1,

$$\gamma_X = \frac{-b_1^2 + b_0 |b_1|}{b_0 b_2 - b_1^2} = \frac{952}{6452}.$$

Hence the smallest positive integer k such that $k \cdot \gamma_X \geq 1$ is 7.

Bibliography

1. Aubin, T.: Equations du type de Monge–Ampére sur les variétés kählériennes compactes. C. R. Acad. Sci. Paris **283**, 119–121 (1976)
2. Bakry, D., Émery, M.: Diffusions hypercontractives. In Séminaire de probabilités, XIX, 1983/84. Lecture Notes in Mathemaatics, vol. 1123, pp. 177–206. Springer, Berlin (1985)
3. Berman, R.J., Boucksom, S., Jonsson, M.: A variational approach to the Yau–Tian–Donaldson conjecture (2018). arXiv (math.DG): 1509.04561
4. Boucksom, S., Hisamoto, T., Jonsson, M.: Uniform K-stability, Duistermaat–Heckman measures and singularities of pairs. Ann. Inst. Fourier **67**, 743–841 (2017)
5. Boucksom, S., Hisamoto, T., Jonsson, M.: Uniform K-stability and asymptotics of energy functionals in Kähler geometry. J. Eur. Math. Soc. **21**, 2905–2944 (2019). arXiv (math.DG):1603.01026
6. Calabi, E.: The space of Kähler metrics. In: Proceedings of the International Congress of Mathematicians, Amsterdam, vol. 2, pp. 206–207 (1954)
7. Calabi, E.: On Kähler manifolds with vanishing canonical class. In: Algebraic Geometry and Topology, A Symposium in Honor of Lefschetz, pp. 78–89. Princeton University Press, Princeton (1955)
8. Calabi, E.: Extremal Kähler metrics. In: Yau, S.T. (ed.) Seminars on Differential Geometry, pp. 259–290. Princeton University Press/University of Tokyo Press, Princeton, New York (1982)
9. Calabi, E.: Extremal Kähler metrics II. In: Chavel, I., Farkas, H.M. (eds.) Differential Geometry and Complex Analysis, pp. 95–114. Springer, Berlin (1985)
10. Chen, X., Cheng, J.: On the constant scalar curvature Kähler metrics, apriori estimates (2017). arXiv. (math.DG) 1712.06697
11. Chen, X., Cheng, J.: On the constant scalar curvature Kähler metrics, existence results (2018). arXiv. (math.DG) 1801.00656
12. Chen, X., Cheng, J.: On the constant scalar curvature Kähler metrics, general automorphism group (2018). arXiv. (math.DG) 1801.05907
13. Chen, X., Donaldson, S., Sun, S.: Kähler–Einstein metrics on Fano manifolds, I: approximation of metrics with cone singularities. J. Am. Math. Soc. **28**, 183–197 (2015)
14. Chen, X., Donaldson, S., Sun, S.: Kähler–Einstein metrics on Fano manifolds, II: limits with cone angle less than 2π. J. Am. Math. Soc. **28**, 199–234 (2015)
15. Chen, X., Donaldson, S., Sun, S.: Kähler–Einstein metrics on Fano manifolds, III: limits as cone angle approaches 2π and completion of the main proof. J. Am. Math. Soc. **28**, 235–278 (2015)
16. Chi, C.-Y.: Pseudonorms and theorems of Torelli type. J. Differ. Geom. **104**, 239–273 (2016)

© The Author(s), under exclusive licence to Springer Nature Singapore Pte Ltd. 2021
T. Mabuchi, *Test Configurations, Stabilities and Canonical Kähler Metrics*,
SpringerBriefs in Mathematics, https://doi.org/10.1007/978-981-16-0500-0

17. Chi, C.-Y., Yau, S.-T.: A geometric approach to problems in birational geometry. Proc. Natl. Acad. Sci. USA **105**, 18696–18701 (2008)
18. Dervan, R.: Uniform stability of twisted constant scalar curvature Kähler metrics (2015). arXiv (math.DG):1412.0648; Int. Math. Res. Not. IMNR, 4728–4783 (2016)
19. Donaldson, S.K.: Scalar curvature and projective embeddings. I. J. Differ. Geom. **59**, 479–522 (2001)
20. Donaldson, S.K.: Scalar curvature and stability of toric varieties. J. Differ. Geom. **62**, 289–349 (2002)
21. Donaldson, S.K.: Lower bounds on the Calabi functional. J. Differ. Geom. **70**, 453–472 (2005)
22. Fujiki, A.: On automorphism groups of compact Kähler manifolds. Invent. Math. **44**, 225–258 (1978)
23. Futaki, A.: An obstruction to the existence of Einstein–Kähler metrics. Invent. Math. **73**, 437–443 (1983)
24. Futaki, A.: Kähler–Einstein Metrics and Integral Invariants. Lecture Notes in Mathematics, vol. 1314. Springer, Heidelberg (1988)
25. Futaki, A.: Asymptotic Chow semi-stability and integral invariants. Int. J. Math. **15**, 967–979 (2004)
26. Futaki, A., Mabuchi, T.: Bilinear forms and extremal Kähler vector fields associated with Kähler classes. Math. Ann. **301**, 199–210 (1995)
27. Guan, Z.-D.: Existence of extremal metrics on compact almost homogeneous Kähler manifolds with two ends. Trans. Am. Math. Soc. **347**, 2255–2262 (1995)
28. Guan, Z.D.: On modified Mabuchi functional and Mabuchi moduli space of Kähler metrics on toric bundles. Math. Res. Lett. **6**, 547–555 (1999)
29. He, W.: On Calabi's extremal metrics and properness (2018). arXiv (math.DG): 1801.07636
30. Hisamoto, T.: On the limit of spectral measures associated to a test configuration of a polarized Kähler manifold (2012). arXiv (math.DG): 1211.2324; J. Reine Angew. Math. **713**, 129–148 (2016)
31. Hisamoto, T.: Orthogonal projection of a test configuration to vector fields (2017). arXiv (math.DG): 1610.07158
32. Hisamoto, T.: Mabuchi's soliton metric and relative D-stability (2020). arXiv (math.DG): 1905.05948
33. Hwang, A.: On existence of Kähler metrics with constant scalar curvature. Osaka J. Math. **31**, 561–595 (1994)
34. Imayoshi, Y., Mabuchi, T.: A Torelli-type theorem for stable curves. In: Geometry and Analysis on Complex Manifolds, pp. 75–95. World Scientific, Singapore (1994)
35. Koiso, N.: On rotationally symmetric Hamilton's equation for Kähler–Einstein metrics. In: Recent Topics in Differential and Analytic Geometry. Advanced Studies in Pure Mathematics, vol. 18-I, pp. 327–337. Academic Press, Boston (1990)
36. Koiso, N., Sakane, Y.: Non-homogeneous Kähler-Einstein metrics on compact complex manifolds, II. Osaka J. Math. **25**, 933–959 (1988)
37. Li, C., Xu, C.: Special test configuration and K-stability of Fano varieties. Ann. Math. **180**, 197–232 (2014)
38. Lichnerowicz, A.: Sur les transformations analytiques des variété kählérienne. C. R. Acad. Sci. Paris **244**, 3011–3014 (1957)
39. Lu, Z.: On the lower terms of the asymptotic expansion of Tian–Yau–Zelditch. Am. J. Math. **122**, 235–273 (2000)
40. Luo, H.: Geometric criterion for Gieseker–Mumford stability of polarized manifolds. J. Differ. Geom. **49**, 577–599 (1998)
41. Li, Y., Zhou, B.: Mabuchi metrics and properness of the modified Ding functional. Pac. J. Math. **302**, 659–692 (2019)
42. Mabuchi, T.: K-energy maps integrating Futaki invariants. Tohoku Math. J. **38**, 575–593 (1986)
43. Mabuchi, T.: Einstein–Kähler forms, Futaki invariants and convex geometry on toric Fano varieties. Osaka J. Math. **24**, 705–737 (1987)

44. Mabuchi, T.: Orthogonality in the geometry of L^p-spaces. In J. Noguchi et al., (eds.) Geometric Complex Analysis, pp. 409–417. World Scientific, Singapore (1996)
45. Mabuchi, T.: Kähler–Einstein metrics for manifolds with non vanishing Futaki character. Tohoku Math. J. **53**, 171–182 (2000)
46. Mabuchi, T.: Vector field energies and critical metrics on Kähler manifolds. Nagoya Math. J. **162**, 41–63 (2001)
47. Mabuchi, T.: Multiplier Hermitian structures on Kähler manifolds. Nagoya Math. J. **170**, 73–115 (2003)
48. Mabuchi, T.: Stability of extremal Kähler manifolds. Osaka J. Math. **41**, 563–582 (2004)
49. Mabuchi, T.: Chow-stability and Hilbert-stability in Mumford's geometric invariant theory. Osaka J. Math. **45**, 833–846 (2008)
50. Mabuchi, T.: Asymptotics of polybalanced metrics under relative stability constraints. Osaka J. Math. **48**, 845–856 (2011)
51. Mabuchi, T.: Relative stability and extremal metrics. J. Math. Soc. Japan **66**, 535–563 (2014)
52. Mabuchi, T.: The Donaldson–Futaki invariant for sequences of test configurations. In: Geometry and Analysis on Manifolds, Progr. Math., vol. 308, pp. 395–403. Birkhäuser, Boston (2015)
53. Mabuchi, T.: The Yau–Tian–Donaldson conjecture for general polarizations, I. In: Geometry and Topology of Manifolds, 10th China-Japan Conference 2014. Springer Proceedings in Mathematics & Statistics, vol. 154, pp. 235–245 (2016).
54. Mabuchi, T.: Asymptotic polybalanced kernels on extremal Kähler manifolds. Asian J. Math. **22**, 647–664 (2018)
55. Mabuchi, T., Nitta, Y.: Strong K-stability and asymptotic Chow stability. In: Geometry and Analysis on Manifolds, Progress in Mathematics, vol. 308, pp. 405–411. Birkhäuser, Boston (2015).
56. Masuda, M., Moser-Jauslin, L., Petrie, T.: The equivariant Serre problem for abelian groups. Topology **35**, 329–334 (1996)
57. Masur, H.: The extension of the Weil-Petersson metric to the boundary of Teichmuller space. Duke Math. J. **43**, 623–635 (1976)
58. Matsushima, Y.: Sur la structure du groupe d'homéomorphismes analytiques d'une certaine variété kählérienne. Nagoya Math. J. **11**, 145–150 (1957)
59. Matsushima, Y.: Espaces homogènes de Stein des groupes de Lie complexes II. Nagoya Math. J. **18**, 153–164 (1961)
60. Moriwaki, A.: The continuity of Deligne's pairing. Int. Math. Res. Not. **19**, 1057–1066 (1999)
61. Mumford, D.: Varieties defined by quadratic equations. In: Questions on Algebraic Varieties (C.I.M.E., III Ciclo, Varenna 1969), pp. 29–100. Cremonese, Rome (1970)
62. Mumford, D.: Stability of projective varieties. L'Enseignement Math. **23**, 39–110 (1977)
63. Mumford, D., Fogarty, J., Kirwan, F.: Geometric Invariant Theory, Ergebnisse der Mathematik und ihrer Grenzgebiete (2), vol. 34, 3rd edn. Springer, Berlin (1994)
64. Nill, B., Paffenholz, A.: Examples of Kähler–Einstein toric Fano manifolds associated to nonsymmetric reflexive polytopes. Beitr. Algebra Geom. **52**, 297–304 (2011)
65. Odaka, Y.: A generalization of the Ross-Thomas slope theory. Osaka J. Math. **50**, 171–185 (2013)
66. Ono, H., Sano, Y., Yotsutani, N.: An example of an asymptotically Chow unstable manifold with constant scalar curvature. Ann. Inst. Fourier **62**, 1265–1287 (2012)
67. Perelman, G.: The entropy formula for the Ricci flow and its geometric applications (2002). arXiv (math.DG): 0211159
68. Phong, D.H., Sturm, J.: Scalar curvature, moment maps and the Deligne pairing (2003). arXiv (math.DG):0209098; Am. J. Math. **126**, 693–712 (2004)
69. Royden, H.L.: Automorphisms and isometries of Teichmüller spaces. In: Advances in the Theory of Riemann Surfaces. Ann. Math. Stud. **66**, 369–383 (1971)
70. Sano, Y.: On stability criterion of complete intersections. J. Geom. Anal. **14**, 533–544 (2004)
71. Sano, Y., Tipler, C.: A moment map picture of relative balanced metrics on extremal Kähler manifolds (2017). arXiv (math.DG): 1703.09458
72. Seyyedali, R.: Relative Chow stability and extremal metrics. Adv. Math. **316**, 770–805 (2017)

73. Simanca, S.R.: A K-energy characterization of extremal Kähler metrics. Proc. Am. Math. Soc. **128**, 1531–1535 (2000)

74. Siu, Y.-T.: The existence of Kähler–Einstein metrics on manifolds with positive anticanonical line bundle and a suitable finite symmetry group. Ann. Math. **127**, 585–627 (1988)

75. Stoppa, J., Székelyhidi, G.: Relative K-stability of extremal metrics. J. Eur. Math. Soc. **13**, 899–909 (2011)

76. Székelyhidi, G.: Extremal metrics and K-stability. Bull. Lond. Math. Soc. **39**, 76–84 (2007)

77. Tian, G.: On a set of polarized Kähler metrics on algebraic manifolds. J. Differ. Geom. **32**, 99–130 (1990)

78. Tian, G.: On Calabi's conjecture for complex surfaces with positive first Chern class. Invent. Math. **101**, 101–172 (1990)

79. Tian, G.: Kähler–Einstein metrics with positive scalar curvature. Invent. Math. **130**, 1–37 (1997)

80. Tian, G.: K-stability and Kähler–Einstein metrics. Commun. Pure Appl. Math. **68**, 1085–1156 (2015); Corrigendum, **68**, 2082–2083 (2015)

81. Tian, G., Yau, S.-T.: Kähler–Einstein metrics on complex surfaces with $C_1 > 0$. Commun. Math. Phys. **112**, 175–203 (1987)

82. Wang, X.: Height and GIT weight. Math. Res. Lett. **19**, 909–926 (2012)

83. Wang, X.-J., Zhu, X.: Kähler–Ricci solitons on toric manifolds with positive first Chern class. Adv. Math. **188**, 87–103 (2004)

84. Wei, G., Wylie, W.: Comparison geometry for the Bakry–Emery Ricci tensor. J. Differ. Geom. **83**, 377–405 (2009)

85. Wu, H.: Old and new invariant metrics on complex manifolds. In: Several Complex Variables (Stockholm, 1987/1988). Math. Notes., vol. 38, pp. 640–682. Princeton University Press, Princeton (1993)

86. Yau, S.-T.: On Calabi's conjecture and some new results in algebraic geometry. Proc. Nat. Acad. USA **74**, 1798–1799 (1977)

87. Yau, S.-T.: On the Ricci curvature of a compact Kähler manifold and the complex Monge–Ampère equation, I. Commun. Pure Appl. Math. **31**, 339–411 (1978)

88. Yau, S.-T.: Open problems in geometry. Proc. Symp. Pure Math. **54**, 1–28 (1993)

89. Yau, S.-T.: On the pseudonorm project of birational classification of algebraic varieties. In: Geometry and Analysis on Manifolds, Progress in Mathematics, vol. 308, pp. 327–339. Birkhäuser, Boston (2015).

90. Yao, Y.: Mabuchi metrics and relative Ding stability of toric Fano varieties (2017). arXiv (math.DG): 1701.04016

91. Yao, Y.: Relative Ding stability and an obstruction to the existence of Mabuchi solitons (2019). arXiv (math.DG): 1908.09518

92. Zelditch, S.: Szegö kernels and a theorem of Tian. Int. Math. Res. Not. **6**, 317–331 (1998)

93. Zhang, S.: Heights and reductions of semi-stable varieties. Compos. Math. **104**, 77–105 (1996)

Printed in the United States
by Baker & Taylor Publisher Services

Printed in the United States
by Baker & Taylor Publisher Services